Conservation Biology Principles for Forested Landscapes

Edited by Joan Voller and Scott Harrison

Conservation Biology Principles
for Forested Landscapes

UBCPress / Vancouver

Printed in Canada on acid-free paper ∞

ISBN 0-7748-0630-3 (hardcover)
ISBN 0-7748-0629-X (paperback)

Canadian Cataloguing in Publication Data

Main entry under title:
Conservation biology principles for forested landscapes

Includes bibliographical references and index.
ISBN 0-7748-0630-3 (bound); ISBN 0-7748-0629-X (pbk.)

    1. Forest ecology.   2. Sustainable forestry.   I. Voller, Joan, 1960–
II. Harrison, Scott, 1962–

QH541.5.F6C66 1998   577.3   C97-911075-0

UBC Press gratefully acknowledges the ongoing support to its publishing program from the Canada Council for the Arts, the British Columbia Arts Council, and the Department of Canadian Heritage of the Government of Canada.

Set in Stone by Irma Rodriguez
Copy editor: Francis J. Chow
Indexer: Annette Lorek
Printed and bound in Canada by Friesens

UBC Press
University of British Columbia
6344 Memorial Road
Vancouver, BC V6T 1Z2
(604) 822-5959
Fax: 1-800-668-0821
E-mail: orders@ubcpress.ubc.ca
http://www.ubcpress.ubc.ca

Contents

Preface

Among the biotic factors that shape the forested landscape, two organisms modify the landscape more than any others: bark beetles (*Dendroctonus* spp.) and humans. Both can affect large spatial scales of a forested landscape on very short time scales. Beetles and humans also follow similar phases in their use of the forest. As they colonize new areas, they use the forest as a store of provisions for basic survival. Bark beetles use the trees for food and shelter. Humans use the trees, other plants, and animals provided by the forest for shelter, fuel, food, and medicine. In both cases, the changes to the landscape are localized as pockets of trees are killed.

Once they satisfy their primary survival requirements, beetle and human populations can begin to grow rapidly. Although beetles and humans respond to different cues, the effects on the forest may be similar. For the beetles, abiotic conditions have important effects on population growth rates. Late fall frosts and mild winter temperatures can enable explosive beetle population outbreaks. When these outbreaks occur, the changes to the forested landscape can be swift and severe. Once humans have attained their minimal survival requirements, they may view the forest as a source of commodities rather than provisions. Commodities are traded or sold to raise the comfort level of daily life. A higher standard of living enables explosive human population outbreaks. When these outbreaks occur, the changes to the forested landscape are swift and severe.

The relationship of bark beetles with the forest does not progress beyond the exploitation phase. For humans, however, the standard of living can potentially rise to a point where their relationship with the forest enters a new phase. Basic survival is no longer an issue, and the appetite for ever more material possessions is sated. In this phase, forested landscapes are valued as habitat for other organisms, for their aesthetic qualities, and as places of solace.

This comparison between bark beetles and humans draws attention to a premise of conservation biology that forms the basis of this book: humans should interact with the forested landscape in a manner that is ecologically sustainable. It is intriguing that we, as long-lived, intellectual organisms, often claim that our relationship with the land is sustainable and then interact with the forested landscape at a spatial and temporal scale similar to that of the short-lived bark beetle.

In this book, we define ecological sustainability as *the perpetual conservation of ecological processes so that the biological productivity of the air, land, and water persists without the use of non-renewable input.* The following are some key points of this definition:

- *Sustainable* means forever and, within the bounds of natural variation, is timeless.
- Forest use is based on *ecology,* not *economics.*
- Conservation of *processes* requires that *rates of change* occur on a time scale that is similar to the natural variation in the system without humans.
- *Knowledge and understanding* of ecological processes and rates are important to their conservation.
- Plans for sustainable use should not require human interventions to 'speed up' ecological processes.

In writing this book, we recognize that humans will continue to use forests and that decisions about land use will continue to be made without perfect information. Our intent is to improve the planning and decision-making processes by providing ecological information on issues of forest use. In doing so, we recognize that the forested landscape is no longer pristine. New approaches must acknowledge the spatial and temporal impacts that have come before. Some current land-use approaches are not working. Where there is information on new, ecologically sustainable approaches, practitioners should switch. Where information on a better approach is not yet available, practitioners should replace the current inappropriate approach with a variety of flexible ones that offer the opportunity to change with new knowledge.

Conservation Biology Principles for Forested Landscapes provides information for those who wish to interact with the land base in an ecologically sustainable manner. It will be useful to students of forestry, ecology, conservation biology, and other natural sciences. We hope that practitioners charged with the administration of land-based programs will make use of the information presented here. We also believe that this book will serve as a resource for the many community groups that are an integral part of all land-use discussions.

Scott Harrison
University of British Columbia
Centre for Applied Conservation Biology

Acknowledgements

We are grateful to all those who contributed their knowledge and time to this project. In particular, Steve Voller and Georgie Harrison provided tireless support. We acknowledge the librarians at the B.C. Ministry of Forests library for their effort. Thanks go to Pat Hedman, Thora Clarkson, and Grace Darling for their time on the computer and in the library, and to Brian Nyberg and Evelyn Hamilton for their assistance with the logistics. Many people reviewed the chapters and added valuable comments. Their efforts and insights are appreciated: Joe Antos, Ralph Archibald, Allen Banner, Shannon Berch, Kim Brunt, Evelyn Bull, Fred Bunnell, Philip Burton, Steve Buskirk, Ann Chan-McLeod, Jiquan Chen, Gerry Davis, Andy Derocher, Matt Fairbarns, Mike Fenger, Dave Fraser, Tony Hamilton, Georgie Harrison, Richard Hobbs, Robin Hoffos, Walt Klenner, Laurie Kremsater, Ken Lertzman, Dennis Lloyd, Todd Manning, Val Marshall, Bill McComb, Greg McKinnon, Scott McNay, Brian Nyberg, Rick Page, Roberta Parish, Les Peterson, Jim Pojar, Chris Ritchie, Stan Rowe, Dennis Saunders, Jean-Pierre Savard, Wilf Schofield, Dale Seip, Doug Steventon, Fred Swanson, Rob Thomson, Peter Tschaplinski, Steve Voller, David Wallin, and Michaela Waterhouse. We would also like to thank the staff at UBC Press for their work in publishing this book.

Conservation Biology Principles for Forested Landscapes

1
Natural Disturbance Ecology
John Parminter

Definitions

Disturbance Ecology

Odum's (1971) definition of an ecosystem included all the organisms in a given area, their interactions with the physical environment, and the creation of a structure with diversity and material cycles. Hann's (1992) more recent definition not only includes the biotic community and its non-living environment but emphasizes the complete range of energy exchange processes – notably plant growth and death, fire, the biotic food chain, weather (average and severe events), the hydrologic cycle, landslides, mass movements, snow avalanches, and the soil-biotic nutrient cycle. The inclusion of these factors reflects the importance of both scale and process and their interactions with time.

Natural communities are dynamic, spatially heterogeneous systems at all scales (Sousa 1984). Whereas spatial variation in ecosystems is obvious, temporal variation may be more difficult to visualize. Ecosystems vary across space in response to both changes in the abiotic environment and the history of natural disturbances. They change through time in response to growth, death, and the effects of chronic low-level disturbances (Hann 1992). Terrestrial ecosystems experience natural disturbances at different spatial and temporal scales; these disturbances are of varying intensities and have varying effects.

Natural disturbances fall into two categories: abiotic and biotic. Abiotic disturbances result from non-living factors such as wildfire, wind, landslides, snow avalanches, volcanoes, flooding, and other weather phenomena. Biotic disturbances result primarily from high populations of insects, other animals, and pathogens, as well as from herbivory (grazing and browsing vertebrates and insects) and predation. The characteristics of these natural disturbance agents combine to define a natural disturbance regime based on the area affected, the return interval, and the magnitude of the disturbances.

The following definitions are adapted from White and Pickett (1985).

Area or size: area disturbed; expressed as area per event or area per time period, or area per event per time period, or total area per disturbance per time period.

Distribution: spatial distribution, including relationships to geographic, topographic, environmental, and community gradients.

Disturbance: any relatively discrete event in time that disrupts ecosystem, community, or population structure, and changes resources, substrate availability, or the physical environment.

Frequency: mean number of events per time period; often used for probability of disturbance when expressed as a decimal fraction of events per year.

Magnitude: either *intensity* (physical force of the event per area per time, such as heat released per area per time period for fire, and wind speed for hurricanes) or *severity* (impact on the organism, community, or ecosystem, such as basal area removed).

Perturbation: a departure (explicitly defined) from a normal state, behaviour, or trajectory (also explicitly defined). It is unlikely that natural ecosystems can be characterized in enough detail to warrant frequent use of this term.

Return interval: inverse of frequency; the mean time between disturbances.

Synergism: effect on the occurrence of other disturbances. For example, drought increases fire intensity, and insect damage increases susceptibility to windstorm damage.

White (1987) described seven general principles regarding natural disturbances and their effects:

- Disturbances occur at a variety of temporal and spatial scales.
- Disturbances affect many levels of biological organization.
- Disturbance regimes vary, both regionally and within one landscape.
- Disturbances overlay environmental gradients, both influencing and being influenced by those gradients.
- Disturbances interact.
- Disturbances may result from feedback between the community state (e.g., successional stage) and vulnerability to disturbance.
- Disturbances produce variability in communities.

Because disturbances operate at different scales (from within a forest stand to the landscape level), their effects range from relatively minor (weakening of an individual) to catastrophic (death of populations and major changes to the ecosystem) (White 1979). Nevertheless, disturbances are not necessarily injurious to ecosystem functioning, even though they may at first appear to be (Averill et al. 1995): 'Recurrence of disturbance and recovery

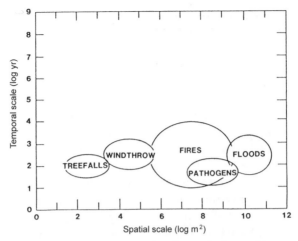

Figure 1.1   Idealized temporal and spatial relationships among disturbance regimes (from Urban et al. 1987).

within ecosystems is an important mechanism for energy flow and nutrient cycling, and for maintaining age, species, genetic, and structural diversity, all attributes of ecosystem health.'

Figure 1.1 shows the temporal and spatial relationships among different disturbance regimes.

Landscape Ecology

The field of landscape ecology integrates natural disturbance regimes and their effects on plant and animal communities. Of interest are the spatial relationships among landscape elements; the flows of energy, nutrients, and species among the elements; and the ecological dynamics of the landscape mosaic through time (Forman 1983). Definitions for specific ecological and landscape characteristics include the following (Forman and Godron 1981; Romme and Knight 1982; Turner 1989; Noss 1990):

Change: alteration in the structure and function of the ecological mosaic through time.

Community: the populations of some or all species coexisting at a site.

Composition: the identity and variety of elements in a collection, including species lists and measures of species and genetic diversity.

Corridor: a narrow strip or linear element that differs from the elements on either side.

Functions: the interactions between the spatial elements, that is, the flow of energy, materials, and organisms among the component ecosystems.

Landscape: a mosaic of heterogeneous landforms, vegetation types, and land uses.

Landscape diversity: the diversity of communities making up the mosaic.

Matrix: the most extensive and contiguous landscape element that exerts the most control over landscape function.

Patch: communities or species assemblages surrounded by a matrix with a dissimilar community structure or composition.

Structure (ecosystem or community): the physical organization or pattern of a system: habitat complexity as measured within communities.

Structure (landscape): the spatial relationships between distinctive ecosystems, that is, the distribution of energy, materials, and species in relation to the sizes, shapes, numbers, kinds, and configurations of components.

Landscape diversity results from two superimposed vegetation patterns: the distribution of species along gradients of limiting factors, and differing patterns of disturbance and recovery within the communities at each point along the environmental gradient (Romme and Knight 1982; Turner 1989). Most landscapes have been modified by humans in one way or another, and therefore consist of a mixture of patches of natural and cultural origin that may be distinctly different from each other in terms of size, shape, and pattern.

A landscape usually consists of patches, corridors, and a background matrix. Disturbances create either a spot (resulting from disturbance of a small area in the matrix) or a remnant patch (an undisturbed remnant of the previous community embedded in a disturbed matrix). Corridors are formed by elongated patches (disturbed areas or remnants) or different habitat types, such as grasslands or riparian zones (see Chapter 3 for more information on corridors). When undisturbed, horizontal landscape structure tends towards homogeneity. A moderate rate of disturbance rapidly increases heterogeneity, and severe disturbance may increase or decrease heterogeneity (Forman and Godron 1981).

Background

Science and Research
Scientific studies in the late 1700s and early 1800s concentrated on describing the geographic distribution of various species of plants and animals. In the early 1900s, F.E. Clements studied the temporal dynamics of vegetation development and popularized a theory of succession that held that a stable endpoint, the climax vegetation, was determined by regional macroclimate (Turner 1989). In 1947, A.S. Watt linked time and space by examining the temporal progression of successional stages and their distribution at a landscape level (Watt 1947).

The prevailing view in this century was that natural disturbances and environmental fluctuations were not important, that the environment was constant, and systems were at equilibrium (Sousa 1984; Sprugel 1991; Reice 1994). The biotic interactions of competition and predator-prey relationships were thought to be the determinants of observed community structure, which would return to its original form after various disturbances. This model is epitomized by the popular concept of the 'balance of nature.' When applied to forests, this 'equilibrium model' argued that natural communities existed in a steady state, sometimes synonymous with climax or old-growth conditions. Protection of old seral stages and avoidance of human intervention were therefore considered the appropriate ways to preserve all species (Oliver 1992).

The 'non-equilibrium model' indicates that community structure is primarily determined by the interactions of a heterogeneous environment, disturbance processes, and the recruitment of individuals and species. Disturbances alone do not determine diversity, but recruitment in a heterogeneously disturbed, patchy environment will result in high overall species diversity. Yet, equilibrium is rarely achieved and the normal condition of most communities and ecosystems is to be recovering from the last disturbance (Johnson and Agee 1988; DeGraaf and Healy 1993; Reice 1994).

Non-equilibrium conditions result when the typical spatial scale of the disturbance approaches or exceeds that of the landscape units, when past events have long-lasting effects on the vegetation, or when the climate is changing (Sprugel 1991). Conversely, a state of dynamic equilibrium can exist when an area is large enough to contain a mosaic of different communities and patches of mature vegetation, various successional stages, and recently disturbed areas (White and Bratton 1980). The best opportunity for equilibrium conditions to develop in the shifting mosaic will occur when disturbances are frequent and at a small scale relative to an otherwise homogeneous landscape, as in the case of canopy gap formation in a large expanse of forest. It is not known whether a landscape is, or has ever been, in equilibrium with a disturbance regime (Pickett and White 1985), but it is clear that spatial scale is an important determinant of the degree of equilibrium that may be present.

Much research has been carried out in North America since the early 1970s, notably on wildfire, to characterize natural disturbance regimes and examine their effects on vegetation succession, stand development, and landscape patterning. Following disturbance, forests regrow through a series of successional stages now called *stand initiation, stem exclusion, understorey reinitiation,* and *old growth* (Oliver 1992). Without further disturbance, succession will continue until a small wildfire or local windthrow more closely resembles its surrounding matrix. Conversely, when a small patch of trees escapes a large wildfire or blowdown, succession will continue until the surrounding disturbed matrix more closely resembles the remnant patch. Both spot disturbances and remnant patches can be relatively short-lived when created by a single natural disturbance, or long-lived when there is repeated disturbance from human activity (Forman and Godron 1981).

The species richness of the plants, mammals, birds, fish, and many invertebrates found in a forest varies with the successional stage of the stand's development. Recognition of the differences among successional stages is important because these define distinct habitats with different suitabilities for plants and animals (Thomas 1979; Harris and Maser 1984; Franklin 1992). Recent studies in the Pacific Northwest have also highlighted the importance of biological legacies (large standing live and dead trees, coarse woody debris, and vegetative propagules) that survive from the pre-disturbance system and influence the speed and direction of ecosystem recovery (Swanson and Franklin 1992).

Because large trees and coarse woody debris are best able to survive an intense large-scale disturbance such as a wildfire, old-growth forests (which have significant quantities of these attributes) are more likely to contribute a greater structural legacy to the future forest than are young or mature forests subject to the same disturbance (Hansen et al. 1991). In coastal Douglas-fir forests in Washington and Oregon, young and old stands had the most coarse woody debris. Young stands had greater numbers of standing dead trees, but old stands had the greatest number of large standing dead trees.

As a consequence of natural disturbances, forests of different composition, age, and structure are distributed across the landscape. Both large and small disturbances have occurred naturally and have perpetuated a state of change rather than one of balance. Large-scale disturbances produce more early successional stages and structures that favour certain vertebrate species. When the interval between disturbances is long and late successional stages are more abundant, along with associated features such as coarse woody debris, other vertebrate species are favoured (Bunnell 1995). To maintain biodiversity, it is important that all successional stages be present on the landscape (Oliver 1995).

Forest Resource Management

British Columbia

In 1937, British Columbia forester F.D. Mulholland described the evolution of forest management practices in many countries as following several well-defined phases. First comes the destructive exploitation of the surplus growth of centuries past, which is considered to be inexhaustible. Then comes protection of the forest crop as the resource becomes valuable and its limits are recognized. Regulation on a sustained-yield basis follows to ensure a secure resource base and stable industry. In the fourth phase, intensive silviculture is added to improve the quality and quantity of the wood supply. Finally, utilization, protection, regulation, and silviculture become part of a process that enables the management of permanent forest reserves for commercial forestry, watershed protection, recreation, and other uses within a socio-economic framework (Mulholland 1937).

Kimmins (1992, 1995) outlined a similar progression, beginning with unregulated forest exploitation and land clearing for other uses. This is followed by administrative forestry, which controls the rate and pattern of forest exploitation to ensure a future supply of forest products and values. However, this second stage is based on legislation and regulation rather than ecological knowledge. A third stage consists of an ecological approach to site assessments and forest practices. This provides for sustainable management of most forest resources but does not necessarily address all of society's concerns regarding forested landscapes, notably stand and landscape aesthetics, spiritual values, and biodiversity. The fourth and final stage of social forestry adds a greater array of resource values on top of environmentally sound forest resource management.

The technologically primitive forest harvesting of the late 1880s was concentrated in the valuable and easily accessible timber on the west coast and the lower elevations of major valleys in the interior of British Columbia. Initial forest management emphasized timber extraction, revenue generation, and rudimentary fire protection activities. Impacts on the forest and other resources were often minimal because only the larger, more valuable and easily transported trees were cut. This produced a canopy gap that was later occupied by the surrounding tree crowns or trees from lower in the canopy.

The use of high-lead logging and extensive railway networks began just before the First World War and affected large areas of land in relatively short time periods, with clearcutting as the harvesting system of choice. Significant reforestation did not take place until the early 1940s because of the prevailing notion that natural regeneration was sufficient by itself. By that time, truck logging had become much more common, and harvesting could be extended to previously inaccessible stands almost anywhere in the province. This had significant implications for forests at higher elevations, and the area affected by harvesting increased substantially after the late 1940s.

Plantation forestry in coastal British Columbia commonly concentrated on Douglas-fir, regardless of site suitability, but was supplemented by natural regeneration of various species. Operational reforestation started in 1939, but lagged far behind in the interior, where it began in 1950. Between then and 1962, reforestation totalled only 4,144 ha in the interior, compared with 65,483 ha on the coast for the same period (Hodgins 1964). Again, natural regeneration was relied upon in most instances.

Snags were felled in accordance with the fire prevention regulations and as a safety measure for planting crews. Secondary streams more often than not received little protection from the effects of logging, and the habitat requirements of wildlife other than big game species were largely ignored.

This picture began to change by the mid-1960s and in the early 1970s, when the Planning Guidelines for Coast Logging Operations (Cameron 1972) addressed environmental concerns. They dealt primarily with wildlife habitat, soil conservation, stream protection, riparian zone management, silvicultural systems, cutblock size and arrangement, and road location. Specific planning systems, such as the resource folio approach, were also instituted in the 1970s, notably in the Prince George Forest Region, to resolve differing resource management objectives and avoid resource conflicts.

With passage of the Ministry of Forests Act on 1 January 1979, the integrated and coordinated management of timber, range, fisheries, wildlife, water, outdoor recreation, and other natural resource values became a legislated objective of the Ministry of Forests (Revised Statutes of British Columbia 1979). Nevertheless, in almost all situations, timber harvesting remained the dominant activity (B.C. Ministry of Forests 1984).

Integrated use involved the alteration of harvesting prescriptions to provide benefits to established resource users, thereby mitigating the effects of harvesting and facilitating change in forest conditions to support new uses. The main variables in harvesting prescriptions were dispersal of activities, size of cutblocks, time interval between the sequential removal of adjacent blocks, and location of roads and other infrastructure (B.C. Ministry of Forests 1984). An additional concern, especially for timber supply planning, was the resulting age structure of the forest expressed in terms of the proportion of mature timber or the average age of the forest.

More recently, specific concerns, management practices, and potential resource use conflicts have been addressed through mechanisms such as the Coastal Fisheries/Forestry Guidelines (B.C. Ministry of Forests et al. 1993). The Forest Practices Code's legislation and regulations as well as the related guidebooks and other documents address a wide variety of resource management issues, planning processes, and operational activities. The evolution of the Forest Practices Code can be traced back to the early 1940s, but the code has its strongest roots in the 1970s and 1980s.

The Pacific Northwest
Because of geographical and ecological connections, there is considerable exchange of ideas regarding forest management practices between British Columbia (at least in the coastal and southern interior regions) and the Pacific Northwest states of Washington, Oregon, Idaho, and Montana. In spite of political, administrative, and cultural differences, there are many parallels in the evolution of forest management policies and approaches in these different jurisdictions.

Forest harvesting in the Pacific Northwest during the late 1800s and early 1900s concentrated on the best timber and put economic concerns ahead of environmental protection. Unmerchantable trees were left standing, slashburning was common, and continuous clearcutting through entire watersheds was frequently carried out (Tappeiner et al. 1997). This approach probably corresponded better with the natural disturbance regime of large wildfires than the staggered pattern of dispersed clearcutting that followed. While partial cutting was tried in Douglas-fir stands, it proved problematic and was all but abandoned by the late 1940s in favour of clearcutting (Tappeiner et all. 1997).

From the 1940s to the 1980s, forest management and research in the Pacific Northwest focused largely on harvesting natural stands and replacing them with Douglas-fir plantations. This was carried out by clearcutting, snag falling, slashburning, suppressing competing vegetation, establishing dense monocultures, and hastening crown closure (Swanson and Berg 1991; Swanson and Franklin 1992). These practices were considered consistent with the prevailing idea that clearcutting and slashburning mimic wildfire effects but are preferable to wildfire because the wood is harvested and utilized rather than wasted.

The forests that resulted from this traditional, intensive plantation approach were structurally and compositionally simpler than natural forests: the early seral stages were much shorter in duration, and fewer stands would live to be older than 100 years. These plantation forests lacked the multilayered canopy, range of tree sizes, and abundant standing dead trees and coarse woody debris that existed in natural forests (Hansen et al. 1991). An illustration of the developmental differences between stands managed for both structural diversity and wood production and stands managed for maximum wood production alone is shown in Figure 1.2

It was thought that these plantations may not be capable of supporting various wildlife species or providing for nitrogen-fixing organisms, thus decreasing site productivity. In addition, a simpler forest may shift predator-prey relations, possibly affecting the role of invertebrates and insectivorous birds as predators of forest insect pests (Swanson and Franklin 1992). More recently, concerns have been expressed and debates held over issues of soil productivity, watershed conditions, and wildlife species that rely on older forest habitat (Swanson and Franklin 1992).

At the landscape level, this management regime involved scattered 8-16 ha cutblocks and resulted in high levels of forest fragmentation (Gillis 1990). The edges of the cutblocks were probably straighter and more abrupt and had a higher contrast than the edges of naturally disturbed patches (Hansen et al. 1991). The breaking up of large blocks of mature forest into a mosaic of young plantations, mature forests, and non-forested land altered disturbance regimes and resulted in lost habitat and reduced habitat quality for some species (Spies et al. 1994). As the proportion of cutblocks in a landscape increases, so does the amount of edge, the degree of isolation of forest

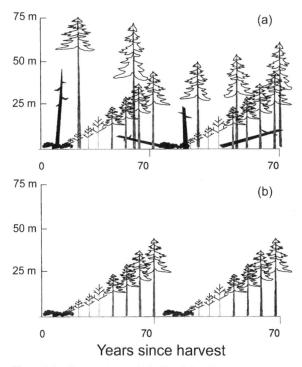

Figure 1.2 Comparison of idealized development
between stands designed for both structural diversity and
wood production (a) and for maximum wood production
only (b) (from Hansen et al. 1991).

remnants, and the length of forest road networks. These factors further influence natural disturbance regimes, often resulting in accelerated windthrow, pest outbreaks, wildfire, and landslides (Franklin and Forman 1987).

The inevitable evolution of forest management practices meant that regulation (avoiding undesirable activities) gave way to sustained-yield management (focusing on a few desired products) and developed into sustainable ecosystem management (for the sake of ecosystem well-being and the production of goods and services). Ecosystem management aims to maintain native species, ecosystem processes, and structures, as well as long-term ecosystem productivity, while also producing wood fibre (Cissel et al. 1994). Under this management approach, forest operations are no longer discrete treatments and the forest is viewed as a functioning ecological system (Swanson and Berg 1991).

With ecosystem management practices, consideration is given to maintaining structural diversity at the stand level and incorporating ecological principles into planning at the landscape level. A premise of landscape ecology is that by integrating human uses with natural landscape-level processes, both the sustainability of human use and viability of protected areas will be increased (Baker 1989).

Recent Developments
Ecosystem management was recently described as consisting of seven major parts or implications (Galindo-Leal and Bunnell 1995). These are that ecosystem management

- is ecosystem-based and that ecosystems are large, heterogeneous, and open
- is practised on a sustainable basis
- appropriately mimics natural disturbance regimes
- aims to maintain viable populations of all native species
- operates at large spatial scales
- plans using long time horizons
- involves interagency coordination and public communication.

The most recent innovation of the ecosystem management approach incorporates natural disturbances in forest planning and operations (Norris et al. 1992; Booth et al. 1993; Grumbine 1994; Kaufmann et al. 1994; Pojar et al. 1994; Stuart-Smith and Hebert 1996) and uses natural ecological processes to form a blueprint for specific resource management activities (Attiwill 1994; Lertzman et al. 1994, 1997). The outcome should be the retention of more natural levels of ecosystem complexity and biodiversity than have resulted from the traditional forest plantation management of previous years.

Where lands are managed to produce various resource-based goods and services, an assumption underlying ecosystem management is that the greater the similarities between the effects of a natural disturbance and the effects of management activities, the greater the probability that natural ecological processes will continue with minimal adverse impact (Rowe 1993): 'The only way to satisfy the popular demand for preserving biodiversity is to practice silviculture and harvesting within large regions in ways that maintain landscape ecosystems in mosaic patterns which approximate or mimic natural mosaic patterns.'

The main concerns in managing protected areas have now shifted to the ecological processes that maintain various species, communities, and ecosystems, rather than focusing on climax communities (Shrader-Frechette and McCoy 1995). Specific questions relate to the extent to which natural disturbances will be permitted to operate in protected areas; the need to accommodate disturbance-dependent species and, conversely, the re-establishment of species lost because of intense local disturbances; the need to manage long-term trends; and boundary issues related to adjacent lands (White and Bratton 1980).

Disturbance Data for British Columbia

In British Columbia, data have been collected on various natural and cultural disturbances since the early part of this century. The total areas affected for the specific time periods of record are summarized in Table 1.1. Figure 1.3 shows the temporal variation for the top three disturbance agents (insects, wildfire, and forest harvesting).

Although these data appear to indicate that the total area affected by natural and cultural disturbances has been rather large compared with the total forested area of the province (54,366,613 ha), not all of these disturbances have resulted in significant forest mortality. Many insects cause only volume growth reductions or deformities in trees, and so the area affected would not necessarily meet the definition of disturbance. Wildfires usually create mosaics of burned and unburned forest, and also occur as less intense surface fires in some ecosystems, burning the understorey vegetation and causing little direct mortality in the overstorey tree canopy.

Other problems with the data are the long time periods, the different methods used to collect information, the mix of agencies involved, and overlaps. For example, a specific area may have had one or more insect infestations and then been harvested later. Many areas were harvested and then slashburned. Others experienced wildfires and then were salvage-logged. Some were burned over by several wildfires this century (this is especially true for surface fires), and some areas have been harvested more than once.

Table 1.1

Areas affected by some natural and cultural disturbances in British Columbia

Disturbance agent	Area (ha)
Insects (1921-95)	24,274,990
Wildfire (1912-96)	10,513,852
Forest harvesting (1912-94)	8,470,340
Slashburning (1913-93)	1,744,789
Land clearing (1913-58)	438,164
Wildlife habitat burning (1982-93)	551,980
Total	46,003,115

Sources: Insect data from various Forest Insect and Disease Survey published reports, Canadian Forest Service, Victoria, B.C.; wildfire and land clearing: annual reports, B.C. Forest Service and Ministry of Forests; forest harvesting: Smith (1981), Spandli (1993), and annual reports, B.C. Forest Service and Ministry of Forests; slashburning and wildlife habitat burning: annual reports, B.C. Forest Service and Ministry of Forests, and annual summaries, Protection Branch, Ministry of Forests, Victoria, B.C.

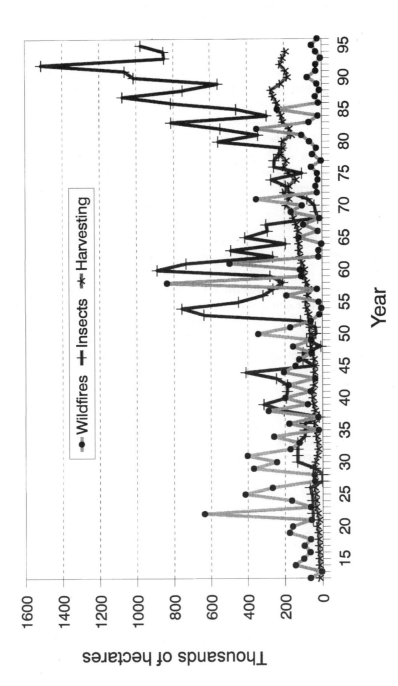

Figure 1.3   Disturbance history of British Columbia's forests with regard to insects, wildfire, and forest harvesting.

These overlaps are counterbalanced by the cultural disturbances that took place but were not noted, including land clearing without fire, since land clearing has been tracked based only on the issuance of burning permits. Significant areas of forested land have been given up for rights-of-way (hydro, transportation corridors, pipelines), urban expansion, and seismic exploration (which may not represent a permanent loss). These factors reduce the total forested area and have a potential influence on the occurrence and propagation of natural disturbances.

There are no long-term quantitative data at the provincial level for windthrow, landslides, snow avalanches, flooding, and weather phenomena such as ice storms and hail. Yet these disturbances have had significant impacts at local and regional scales in historic times as well as over a longer time frame. At its greatest extent, windthrow can affect many thousands of hectares of forest at a time – in single or clustered locations – but it is difficult to quantify when mortality is only partial. Forest diseases, quantifiable at some scales, are difficult to assess provincially, given the endemic nature of many agents.

While much variation exists in the way different disturbance agents operate, and although the data are incomplete, the total area affected by disturbances such as wildfire and forest harvesting is still quite significant. The area harvested since 1913 is estimated to be over 8 million hectares. This represents 19% of the potential commercial forest area of the province. It is important to remember the time scale involved and to realize that most of the harvested area has reforested. The oldest of these new stands are approaching maturity, or are already mature. As mentioned earlier, forest harvesting in the early 1900s was concentrated in more accessible areas and at lower elevations, so the impact on forest age class distributions, for example, has been significant in these locations.

Provincially, the annual area harvested surpassed 100,000 ha for the first time in 1961, and has for the most part been greater than 200,000 ha since the late 1970s. Technology has given us the capability to affect the landscape in all parts of the province, and by using this capability we have altered the landscape's structure and function for a long time to come. There may or may not be negative impacts as a result, but the long-term implications of such alterations need to be examined and evaluated as part of the resource management planning process. Besides altering the landscape that became part of the working forest, we have also suppressed natural disturbances such as wildfires and insect outbreaks in the surrounding matrix. Even if a particular area has not been subjected to resource utilization, it is possible that the landscape's characteristics have been affected by human interference in one way or another.

Ecological Principles

Natural Disturbance Effects

In managing to conserve biodiversity and maintain relatively natural landscapes, the variability in spatial and temporal heterogeneity should be kept within the range that existed as the current ecosystems and landscapes developed (Hann 1992). This raises some potentially unanswerable questions, given the possible range of historical natural variability and our inability to know its every detail. Natural disturbances form cyclic patterns that are not necessarily regular or in equilibrium with a region's climate, topography, and biota.

Disturbance regimes vary at different scales and are relative to specific locations and time intervals. Some locales may be more subject to wind, landslides, and flooding, whereas others are affected more by fire, insects, and diseases. However, both small- and large-scale disturbances due to different causal agents can operate simultaneously in the same community or on the same landscape as a function of local climate, topography, and biota (Pickett and Thompson 1978).

Small-Scale Disturbances

Within a stand, small-scale disturbances primarily involve tree death or treefall and subsequent canopy gap formation. Such gaps result when one to several large trees in the upper canopy, perhaps affected by insects, disease, a structural abnormality, snow loading or glaze ice accretion, flooding, fire, or lightning, die and/or fall over (Sousa 1984; Franklin et al. 1987). Wind can affect both relatively healthy and weakened trees and can result in a windsnapped bole or a windthrown tree with roots attached (Bormann and Likens 1979a).

This local disturbance reduces net primary productivity, increases coarse woody debris loading, adds more water and nutrients to the soil, and speeds up decomposition and mineralization in the canopy gap (Bormann and Likens 1979b). Small-scale disturbances that create canopy gaps are a primary source of heterogeneity in forest structure and composition, enabling species to become established that could not do so under a closed canopy (Lertzman and Krebs 1991). On the west coast of North America, they shorten the time required by a stand to obtain old-growth status by increasing the proportion of attributes associated with old-growth conditions – coarse woody debris, vertical and horizontal heterogeneity, and the proportion of shade-tolerant tree species (Spies and Franklin 1988).

The size and intensity of the local disturbance resulting from tree death or treefall are a function of the number and biomass of the tree(s) that fall.

When a single standing tree dies and gradually deteriorates, the gap will be small. When larger individuals fall, they often strip limbs from, snap off, crush, or uproot smaller trees as they topple to the ground. If that happens, an area of several hundred square metres can potentially be affected by the treefall disturbance (Bormann and Likens 1979b). More than 15% of the mortality in mature and old-growth coastal Douglas-fir forests results from such multiple treefalls (Franklin et al. 1987). Large gaps are also formed where a new gap is adjacent to or overlaps an older one.

In canopy gap formation, variability is ever-present. The gap size typically ranges from 50 to 200 m^2 for single treefalls and from 300 to 500 m^2 for multiple treefalls (Attiwill 1994). One study in the Appalachian Mountains of the eastern United States determined that the largest treefall gap was 1,490 m^2 even though the average was only 31 m^2 (Sousa 1984). In two coastal Oregon forests of Sitka spruce and western hemlock, the medians and ranges for canopy gap sizes were 233 m^2 (12-851 m^2) and 181 m^2 (16-695 m^2), respectively (Taylor 1990). On the lower B.C. coast, a subalpine forest of amabilis fir, western hemlock, mountain hemlock, and yellow-cedar had a median gap size of 41 m^2, while the range was from 5 to 525 m^2 (Lertzman and Krebs 1991).

Blowdown pocket, near Kispiox, B.C. This small gap created by wind will eventually allow new plant species to establish on the forest floor. (Photograph by John Parminter)

Interactions between windthrow, insect outbreaks, and wildfire are common and very likely widespread in the coniferous forests of western North America (Veblen et al. 1994). In subalpine forests of the Rocky Mountains, the spruce beetle (*Dendroctonus rufipennis*) can kill individual trees or patches, and subsequent wildfires can burn some of the beetle-disturbed forests, producing large areas of standing dead trees and open ground. Windthrow may then fell some of these standing dead trees, and the end result is a complex mosaic of patches with different degrees and sequences of disturbance (Baker and Veblen 1990).

Outbreaks of insects such as the Douglas-fir beetle (*Dendroctonus pseudotsugae*) create clustered or widely dispersed small groups of dead trees. This insect is often associated with wind-damaged trees (Spies and Franklin 1988). Windthrown dead and dying trees may support populations of bark beetles and allow them to gain a foothold in nearby healthy trees. Pathogens such as root rots result in patchy mortality, sometimes in all stand age classes and at a regional scale. Many windthrown coastal, old-growth Douglas-fir are infected with the butt rot fungus, *Polyporus schweinitzii* (Franklin et al. 1987).

Reduced fire frequency can change successional processes and alter stand structure and composition. In Douglas-fir and ponderosa pine forests of western Montana, these changes have increased the duration and intensity, but not the frequency, of western spruce budworm (*Choristoneura occidentalis*) outbreaks in this century (Anderson et al. 1987).

Synergistic interactions between insects and disease are common. Disease-weakened trees may be predisposed to insect attack, and insect attack may predispose trees to disease infection (Fellin 1980). Many insects, such as bark beetles and wood borers, disperse a variety of decay fungi. Insects and diseases influence and determine stand composition and structure as well as the direction and rate of succession (van der Kamp 1991; Haack and Byler 1993). They also form vital links in food chains and facilitate other ecological processes, sometimes indirectly creating wildlife habitat, such as when woodpeckers preferentially select trees with heart rot for nest excavation.

In some instances three agents interact with synergistic effects to determine stand dynamics. Lodgepole pine weakened by *Polyporus schweinitzii* are attacked by the mountain pine beetle (*Dendroctonus ponderosae*). The subsequent accumulation of dead fuels increases the probability of a forest fire. The fire leaves basal scars on surviving trees, which are then infected by the fungus – and the process begins again (Geiszler et al. 1980). In a general sense, trees killed by insects may increase the fire hazard if fuels accumulate in sufficient quantities. Fire-killed trees or those weakened by fire damage may be attacked by bark beetles or phloem feeders (Fellin 1980).

Conversely, the effects of one disturbance agent may impart some protection against another agent, albeit indirectly and for a limited time. For example, spruce trees regenerating after wildfire will be immune to spruce beetle attack until they have attained sufficient diameter. In some ecosystems this could be for as long as 70 years. The same is true for sites frequently disturbed by snow avalanches, which also create effective fuelbreaks that limit the spread of wildfires (Veblen et al. 1994).

Lodgepole pine forests regenerating after wildfire will be immune to the mountain pine beetle for about eight decades, but afterwards can be killed by the insect over extensive areas (Safranyik et al. 1974; Furniss and Carolin 1980). The continued interaction between mountain pine beetle mortality, fuel build-up, and extensive wildfire essentially perpetuates lodgepole pine at a landscape scale. In such cases some degree of fragmentation would be beneficial as a preventive measure against extensive muntain pine beetle epidemics (Perry 1988).

Large-Scale Disturbances
Wildfire, wind, landslides, snow avalanches, flooding, and certain other weather phenomena can be of great magnitude and act over large areas. Such catastrophic disturbances affect both healthy and weakened trees and usually result in significant or complete mortality over wide areas. Large-scale disturbances such as wildfire generally return a stand to an earlier seral stage by killing many plants, thereby favouring the establishment of early seral species. On the other hand, windthrown forests may be accelerated towards a later seral stage if shade-tolerant advanced regeneration forms the bulk of the next stand (Spies and Franklin 1988).

Wildfires range from approximating the size of a canopy gap to covering hundreds of thousands of hectares or more, notably in boreal forests. Crown fires create a mosaic of stand conditions by initiating successional processes where the overstorey is killed, leaving remnant patches unburned and partially burning others (Oliver 1995). In riparian areas, topography and fuel build-up can contribute to burns that kill all the trees; at the same time, however, riparian areas that are wetter and/or have fewer fuels may burn less completely, depending on fire weather and fire behaviour.

Historically, surface fires were common in interior Douglas-fir and ponderosa pine forests, and maintained them in an all-aged condition. Normally these forests consisted of distinct groups of similarly aged trees, with an open understorey and interspersed grasslands (Weaver 1974; Wright and Bailey 1982). Such stand-maintaining fires suspend forest succession, with conversion to later successional types being avoided.

Wind damage, more specifically windthrow and windsnap, covers an areal range similar to that of wildfire – from small gaps to landscape scales. Variation in impacts are due to meteorological conditions (wind speed,

Burnt slopes in northern British Columbia. Wildfire favours the establishment of early seral species, often over wide areas. (Photograph by John Parminter)

gustiness, storm duration, soil moisture, snow and rain loading in the tree crowns), topographic characteristics (wind exposure, wind direction, speed, and turbulence), stand and tree characteristics (species composition, stand height and density, crown and bole condition, rooting strength), and soil characteristics (depth, drainage, and relationship to root strength) (Stathers et al. 1994). Extremely large disturbed areas and/or widespread scattered blowdowns result from hurricane-force storms that cause damage regardless of topographic location, tree species, stand characteristics, or soil type (Weetman 1957).

In the Pacific Northwest, strong winds have had variable effects, ranging from scattered blowdowns totalling 30,000 ha (6 December 1906 on Vancouver Island and the Lower Mainland coast) to contiguous areas of up to 544,000 ha (29 January 1921 on the Olympic Peninsula of Washington). In the latter case, Guie (1921) estimated that one-third of the trees in the affected area were blown down, ranging from 25% to 75% in distinct portions. Many of these forests were old growth. The same storm affected the neighbouring west coast of Vancouver Island much less severely, with only 0.25% of the forest of over 2 million hectares experiencing blowdown, often in swaths 100-300 m wide in the valleys (Andrews 1921; Anonymous 1921). In addition to the more localized and variable impacts of individual storms, it is known that tree mortality due to wind events varies regionally, ranging from up to 80% in coastal Sitka spruce–western hemlock forests to less than 15% in interior ponderosa pine forests in the Pacific Northwest (Franklin et al. 1987).

Potential consequences of landslides and flooding include major changes to the structure of surficial materials and drainage channel systems as well as possible off-site impacts. In nearly all cases, a similar ecosystem eventually develops on the site.

Interactions between the abiotic disturbances of wind and wildfire and the biotic disturbances of diseases and insects occur at a large scale as well. Although blowdown areas will have the same fuel loading as the pre-disturbance forest, the changes to their susceptibility to fire ignition due to fuel rearrangement have not yet been rigorously documented. Wildfires of up to 32,000 ha in size have occurred in blowdown areas as early as one year later (Flannigan et al. 1989).

All known major outbreaks of the Douglas-fir beetle in western Washington and Oregon were triggered by severe forest disturbances, notably blowdown as a result of extensive storms and large forest fires (Wright and Harvey 1967). The most serious infestations of Douglas-fir beetle there followed catastrophic windstorms in the mid-1950s and in 1962 (Perry 1988). Even without the predisposing influence of wind, insects can cause extensive tree mortality. This occurred in Colorado in the mid-1800s and the 1940s, when the spruce beetle killed nearly all mature spruce trees over hundreds of thousands of hectares along a 1,200 km axis (Baker and Veblen 1990; Veblen et al. 1994).

The distinctions between small-scale and large-scale disturbances are somewhat artificial if we consider the development and cyclic renewal of terrestrial ecosystems to be intimately linked to disturbances of both kinds at all temporal and spatial scales. It is logical to consider small-scale and large-scale disturbances as endpoints on a continuum, but the more difficult question is really how to define what is normal in a particular system (Bormann and Likens 1979b; White and Pickett 1985). The relationships of disturbances to community dynamics and their roles in maintaining species diversity have been the subject of extensive reviews (Vogl 1980; Sousa 1984; Petraitis et al. 1989; Oliver and Larson 1990; Perry and Amaranthus 1997).

Biodiversity and Natural Disturbances

Variations in geographic location, topography (elevation, slope, and aspect), and climate not only influence current and potential natural vegetation but also help determine the presence or absence of natural disturbance agents. Indeed, the spread of disturbances across a landscape is influenced by spatial heterogeneity (Turner 1989; Reice 1994). Disturbances may be able to spread only within a particular community, or they may cross boundaries and spread from one to another. In the first case, high levels of landscape diversity should retard the further spread of the disturbance. When the disturbance is capable of moving across boundaries and into different communities, its spread may very well be enhanced by landscape heterogeneity (Turner 1989).

The result of the interaction of various disturbance regimes with an already diverse physical and biological environment is the creation and maintenance of ecosystem diversity. It has been argued that definitions of biodiversity must include processes such as interspecific interactions, natural disturbances, and nutrient cycles because they are crucial to maintaining biodiversity (Noss 1990). Natural disturbances and those caused by human actions can promote plant and animal diversity by influencing the composition, age, size, edge characteristics, and distribution of stands across the landscape. Nevertheless, this needs to be examined closely, because too great a diversity of patch types and sizes can fragment the matrix.

Structural complexity provides the basis for much of the variety and richness of species, habitats, and processes (Franklin 1992). The important structural attributes include the size of standing live and dead trees, the condition of those trees, and the size, amount, and condition of coarse woody debris on the forest floor (refer to Chapter 7 for more information on wildlife trees and coarse woody debris). The canopies and boles of standing trees are important habitats for epiphytic organisms and invertebrates. Coarse woody debris provides habitat and a long-term source of nutrients, and may be a site for nitrogen fixation. It also fulfils many important roles in stream ecosystems by forming pools and backwaters, providing nutrients, dissipating the energy of flowing water, and trapping sediment (Sedell et al.

As seen in this photograph taken near the Spatsizi River in northern British Columbia, natural disturbances such as fire can promote biodiversity across a landscape. (Photograph by John Parminter)

1988). All of these are affected by natural disturbances, both small-scale and large-scale.

The stand age profile is primarily determined by the time since disturbance, effects of the disturbance on the site, plant species regeneration mechanisms, post-disturbance environmental conditions, and the time during which plant species regenerate in response to disturbance. Continuing small-scale disturbances such as treefalls will lead to an all-aged profile if shade-tolerant species can occupy the understorey. Canopy gaps created by treefalls will be filled in by crown extension into the gap and/or the establishment and growth of trees in the gap, some of which may have been advance regeneration that was present before the treefall.

In contrast, after a crown fire kills many trees (in patches of various sizes), a fire-adapted species responds by releasing stored seeds or by vegetative reproduction. Additional seeds may be dispersed from individual survivors or remnant patches and from adjacent unburnt stands. The post-fire stand will be quite even-aged, and most trees will probably become established within a few years. In boreal forest types, the early to middle stages of succession can be the most floristically and structurally diverse, with hardwoods and conifers mixed together in a multilayered canopy. As succession proceeds and the seral species such as trembling aspen, paper birch, and lodgepole pine drop out, the stand becomes less diverse and consists primarily of white or black spruce (Kelsall et al. 1977; Foote 1983).

The size of the burnt patch and its edge characteristics are determined by the fire's behaviour, which is a function of the number of ignition points and fuel quantities, arrangements, and availability for burning (due to moisture content), as well as topography, terrain features, wind speed, and wind direction (determinants of the fire's direction of travel and rate of spread). At one end of the spectrum, a fire can create a matrix of burnt patches within a larger unburnt landscape, or it can do the opposite – leave a few unburnt patches within a larger burnt landscape. Because of the number of factors involved, the landscape-level effects of fire are highly variable.

All weather factors being equal, the probability of a fire igniting and spreading depends on the amount, flammability, and distribution (horizontal and vertical) of fuels (Perry 1988). A fire's growth may be limited if adjacent areas have higher fuel moisture contents and/or if the characteristics of the terrain and topography limit fire spread. Landscapes can develop characteristics that reinforce a shorter fire return interval in certain vegetative types that, because of their composition and structure, are more flammable than neighbouring types (Quirk and Sykes 1971).

Later in the fire season, and especially after a period of extended drought, more vegetated communities will contain adequate quantities of available (burnable) fuel and can potentially support a wildfire. In such situations, fire can more easily burn through different vegetative types, although differential flammability and preferential burning may persist as a function of tree species composition, stand age, and fuel arrangement (Perry 1988). In some locations the presence of a stationary high-pressure system serves to dry the fuels, and then, as low-pressure systems intrude, lightning and erratic winds result in fire ignitions and rapid spread (Johnson et al. 1990).

In a similar fashion, a very severe windstorm can blow down stands with different composition and age or different size and edge characteristics, but most affected stands will be older and taller rather than younger and shorter (Sousa 1984). In the latter case, trees tend to have more resilient boles that can tolerate repeated wind stress.

Post-windstorm diversity will be maintained insofar as some surviving individuals and vegetative propagules on the site re-establish a degree of the patch's previous identity and thus counteract the landscape homogenization that would result from widespread establishment of early seral species. Some shift towards late seral species could occur when advanced regeneration forms the bulk of the post-disturbance stand. With wildfire, if the pre-fire forest stands were of different ages, the initial result of large wildfires burning through different types will be a decrease in landscape diversity, because the post-fire stands will be of the same age and exhibit a simpler structure.

The intermediate disturbance hypothesis links localized catastrophic mortality to the maintenance of species diversity and predicts that the highest diversity will occur when there is an intermediate level of disturbance. Diversity results from a balance between the frequency of disturbances that provide recolonization opportunities for species and the rate of competitive exclusion that determines the rate of species extinctions within patches (Petraitis et al. 1989).

If the disturbance is too mild or too rare, patches will be dominated by the few species that are able to outcompete all others. If the disturbance is too harsh or too common, only those species able to resist the disruption will persist. It cannot be said that disturbances always increase landscape diversity. If the severity of the disturbance is low enough to limit its effects to small areas, or if the severity is especially high, decreased diversity is likely to result.

The Need for Management that Includes Natural Disturbance Regimes
In natural unmodified landscapes, disturbances vary in time and space and maintain many seral stages and community types at a regional scale (Noss 1983). Although natural disturbances are common occurrences, habitat fragmentation is especially increased when cultural disturbances dominate the landscape (Probst and Crow 1991). Fragmentation can isolate some members of a population and thereby increase the probability of local extinctions. Fragmentation also affects area-sensitive species by reducing the amount of

available interior habitat, and adversely affects edge-sensitive species when the contrast between adjacent patches is increased (for more information on interior habitat and edges, refer to Chapters 5 and 8.)

Although portions of landscapes will be within protected areas where natural disturbances will continue to varying degrees, the conservation and enhancement of biodiversity must also take place in other wildlands and in managed forests. The protected areas and other lands are embedded in a semi-natural matrix of 'culturally modified' (Noss 1983) forests, grasslands, and wetlands that are subject to a variety of human uses, ranging from concentrated and intensive to dispersed and extensive. Rowe (1992) defined three distinct ecosystems: (1) artificial, which are subject to intensive single-use resource extraction 'farming'; (2) natural, which are preserved as parks and wildernesses; and (3) semi-natural, which are subject to the extraction of multiple resources in perpetuity.

The semi-natural matrix is dominant at the provincial scale, occupies many of the most productive locations, and contains most of the biological diversity. Thus, it plays three vital roles: (1) providing habitat at smaller scales; (2) buffering and increasing the effectiveness of reserved areas; and (3) controlling connectivity in the landscape, including the movement of organisms between protected areas (Franklin 1993).

This photograph taken near the Dease River in northern British Columbia shows landscape patterns created by wildfire. Responsible management is needed to provide similar seral stages and landscape patterns required to support a range of plant and animal species without over-fragmenting the landscape. (Photograph by John Parminter)

Methods for Including Natural Processes in Resource Management
Active management of the landscape matrix is needed to provide the seral
stages and landscape patterns required to support a range of plant and animal species. This is accomplished by the following (Probst and Crow 1991;
Wiedenmann 1992):

- using a regional perspective but also considering within-stand diversity
- thinking beyond the boundaries of specific ownerships when planning
  and managing, and doing so over large areas, not on a stand-by-stand
  basis
- considering the cumulative impact of individual projects on regional
  populations and resources
- emphasizing multispecies and ecosystem management instead of single-
  species and tree management
- emulating the natural disturbance patterns and recovery strategies
- maintaining a diverse mix of genes, species, biological communities, and
  regional ecosystems
- including the full spectrum of seral stages on the landscape, from early
  successional to old (this acts as the 'coarse filter' for genetic and ecosystem diversity)
- minimizing fragmentation and maintaining connectivity
- prioritizing in favour of the species, communities, or processes that are
  endangered or that otherwise warrant special attention
- maintaining or creating spatial patterns (large patches, landscape linkages, and low contrast between adjacent patches) that enhance conditions for problem species
- leaving a mixture of tree sizes and species on the site and restoring naturally diverse forests after harvest
- selecting what you leave behind as carefully as what you take away, especially with regard to the biological legacies of live and dead standing trees
  and coarse woody debris
- conducting ecological surveys and monitoring; designing, implementing, and evaluating new approaches.

Biodiversity includes genetic, species, and ecosystem elements that can
operate at different spatial and temporal scales. In including biodiversity as
an objective of management, it is important to follow an ecosystem management approach at landscape to regional levels rather than a single-species
approach (Probst and Crow 1991; Oliver 1992; Rowe 1993). This avoids the
problems that result when different species in the same area have different
habitat requirements, and it compensates for our lack of knowledge about
species presence and population levels. It also allows for the preservation of

ecological processes in poorly known or unknown habitats and ecological subsystems (Franklin 1993), and provides some insurance against future uncertainties (Burton et al. 1992).

While a single-species approach may be necessary when dealing with an endangered species that requires special consideration, management for species richness at the local level may have inherent problems. This approach can result in higher numbers of habitat generalist organisms at the expense of habitat specialists whose needs can be met by only a limited number of ecosystems. For example, this is one of the dangers confronting the 75 vertebrate species that, as habitat specialists, are closely associated with or dependent upon late successional Douglas-fir forests in the Pacific Northwest (Carey 1989). Similarly, a landscape approach is vital to incorporating the needs of species that utilize multiple habitat types; it includes ecotones as well as species assemblages that change gradually along environmental gradients (Noss 1990).

Current landscape design methodologies also incorporate site history, natural disturbance regimes, and successional processes (Diaz and Apostol 1992; Bell 1994; Diaz and Bell 1997). By paying attention to the patterns resulting from natural disturbances and the terrain in which they occurred, and by anticipating future disturbances, it is possible to integrate management activities into the natural landscape.

In British Columbia the biogeoclimatic ecosystem classification system (Pojar et al. 1987; MacKinnon et al. 1992) provides a regional, local, and chronological framework that can be used to study the historical roles of natural disturbances in different ecosystems, their importance, and their interaction with resource management activities. The more recently developed ecosystem mapping methodology integrates abiotic and biotic ecosystem components and provides basic information about ecosystem distribution, historical site conditions, and diversity at the ecosystem and landscape scales (Banner et al. 1996).

Cissel et al. (1994) described a six-phase process used for a case study that produced a list of potential management actions:

1   assess historical and current disturbance regimes for terrestrial and aquatic systems
2   integrate that information and define a desired landscape condition and associated management approach for subareas that have similar disturbance regimes, potential vegetation, and human-use patterns
3   project the resulting management approach into the future using simple modelling in a geographic information system, assume no natural disturbances but allow for natural succession, and model harvesting that approximates the natural disturbance regime

4   analyze the landscape pattern projected by phase 3 to determine whether adjustments are needed to meet established management objectives (current conditions may be outside the range of desired conditions)

5   make adjustments to alter the frequency, intensity, or location of future harvesting units; change the amount or shape of reserves; or prescribe ecosystem restoration practices

6   compare existing conditions with the planned trajectory of conditions to identify management actions that will encourage desired conditions.

Further steps were taken for the area analyzed. They included examining the ecological and social objectives of alternative management scenarios, increasing public participation in ecosystem management planning, analyzing the context of the study area in relation to its setting, and conducting landscape experiments and modelling to improve understanding of ecosystems at larger scales (Cissel et al. 1994).

At the stand level, the concept of retaining structural and compositional diversity is based on, and requires an improved understanding of, the following (Swanson and Berg 1991):

• disturbance processes in natural and managed forests
• ecosystem functioning
• experience with managed forests, including successes and failures in implementing intensive forestry practices.

Research has shown that standing dead trees and coarse woody debris represent habitat for many species and are needed to provide for biodiversity and forest productivity (Swanson and Franklin 1992). Traditional intensive plantation forestry greatly reduces the input of coarse woody debris after the residual material from the previous natural stand has decomposed (Spies et al. 1988). In comparing the ecological effects of a wildfire with traditional clearcutting, it is apparent that harvesting removes the live wood (reducing coarse woody debris loading and structural diversity); creates a pattern of smaller, dispersed, and more uniform patches (affecting wildlife habitat and future disturbance probabilities); and truncates succession before it can proceed to older stages (decreasing the quantity of stands with old-growth characteristics) (Spies and Franklin 1988).

Taking our cues from studies of the effects of natural disturbances and the value of biological legacies, new practices can be implemented in stands with any history either at the initial cutting stage, when carrying out thinnings, or as part of salvage logging in windthrown or burnt areas. Some of the components that are retained are standing dead and live trees, coarse woody debris, and riparian habitats. The long-term consequences of this new management regime need to be evaluated over several cutting cycles and at a landscape scale (Swanson and Franklin 1992). Management activities influence the landscape pattern not only by harvesting but also through management of natural disturbances and designation of protected areas and a variety of other reserves.

The focus should be on the effects of natural processes and disturbances rather than on the particular agent responsible. For example, a canopy gap can result from the death of a single tree or a group of trees, and be caused by senescence, an abiotic agent, a biotic agent, or some combination of these. If the natural disturbance regime usually operates at a within-stand level, management activities could emulate the canopy gap formation process through single-tree or group selection silvicultural systems. The site's characteristics suitability for treatment will have to be considered in terms of how it might affect soil processes; understorey plant and tree regeneration, survival, and growth; wildlife and their habitat requirements; forest insects and diseases; fuel complexes and fire hazard; and windthrow susceptibility.

Larger-scale disturbances such as more extensive wildfire and windthrow can be emulated at the landscape level by designing a cutblock of similar size and shape, leaving remnant patches of live and dead trees as well as residual coarse woody debris. Aggregations of cutblocks may be needed to emulate larger natural disturbances, especially wildfires in sub-boreal and boreal ecosystems. At a landscape scale, the choice of rotation age, rate of cut, and cutblock layout will determine the future age class distribution and landscape pattern.

Because of our limited understanding of natural disturbance regimes, management that aspires to emulate natural disturbances has so far generally focused on wildfire and windthrow. It would be difficult and counterproductive to emulate landslides and severe flooding. Insects and diseases are problematic as models because they are both endemic and epidemic, do not always cause mortality, and operate at various spatial and temporal scales. Despite this variation, however, they can serve as management models to some extent.

Scale Issues, Variation, and Predictability
Natural disturbances maintain structure at the species, ecosystem, and landscape scales. What constitutes a disturbance as opposed to a normal fluctuation depends on the scale of observation and on the level of biological organization being considered (Baker 1992). Natural disturbances also vary on a continental scale as a reflection of spatial variations in climatic conditions, vegetation types, and physical settings. For practical purposes one can use the hierarchy described by Urban et al. (1987):

Level	Boundary definition	Scale
Landscape	Physiographic provinces Changes in land use or disturbance regimes	10,000 ha
Watershed	Local drainage basins Topographic divides	100 to 1,000 ha
Stand	Topographic positions Disturbance patches	1 to 10 ha
Gap	Large-tree influences	0.01 to 0.1 ha

Interactions at one level will determine the behaviour of a component at the next higher level. Each level of the hierarchy contains the levels found below it. The size of individual disturbance events and any larger-scale disturbance mosaic that results will determine the appropriate scale at which to study the disturbance and its effects.

Natural disturbances can occur over a broad range of patch sizes, with different irregular shapes. They can be of varying intensities (little to all trees or dead wood remaining), with clumped to scattered patterns and different frequencies (canopy gaps can occur yearly but stand-replacing fires may occur in the same stand only once in 200 years). By contrast,

Natural disturbances such as wildfire can occur over a broad range of patch sizes and can be of varying intensities and differing frequencies. (Photograph by John Parminter)

traditional forest management in the Pacific Northwest utilized a narrow range of patch sizes (4-40 ha), shapes (quite regular), intensities (few residual trees or dead wood left), and frequencies (60-120 years) on a scattered landscape pattern (McComb et al. 1994). It is in recognition of these differences that resource managers now seek to incorporate knowledge of the effects of natural disturbances into management prescriptions. To accommodate the effects of future natural disturbances, it is also advisable to base management actions at least in part on the natural disturbance regime of each site (Swanson and Franklin 1992).

However, the range of natural variation that occurs soon leads to difficulties in the study of natural disturbances. A lengthy time without disturbance will allow landscape patches to become comparatively old, whereas frequent disturbances will result in more young landscape patches, even if the size of individual disturbances does not change (Baker 1992). Conversely, landscape structure will fluctuate through time when the frequency is constant and the disturbance area changes randomly. The fluctuations will be greater when unequal-sized disturbances occur at unequal time intervals.

It is important to understand the dynamics of natural disturbances, both those that are most frequent and those that are rarer and possibly extensive. The importance of the latter may not be obvious in contemporary time scales, but they may be critical to the structure, function, and pattern of natural landscapes (Pickett and Thompson 1978; White 1987). At the same time, especially in the context of management of protected areas, questions may be raised regarding the potential detrimental effects of natural disturbances, as they can have a significant impact on species or processes (Turner et al. 1994).

Ultimately, it may be impossible to accurately define the natural vegetation or the natural disturbance regime in many areas. Even without the impact of human activities, climatic variability, changing conditions, and the passage of time cause dramatic changes in climax vegetation and disturbance regimes. In an environment not in equilibrium, the notion of having a unique vegetation or disturbance regime may be flawed because the range of natural vegetation and associated processes is probably much greater than commonly imagined (Sprugel 1991).

Nevertheless, the range of natural disturbance regimes can be interpreted from dendrochronological, paleoecological, and archival studies and displayed on maps showing units of disturbance frequency and severity. The historical range of variability concept was developed to describe the dynamics of ecosystems that are undergoing continual change. It allows us to identify the range of future conditions that are sustainable and to establish the limits of acceptable change (Morgan et al. 1994). Understanding the range of natural variability enables us to design management prescriptions and provides a reference point for evaluating the success of ecosystem management (Swanson et al. 1994).

No single silvicultural system will precisely emulate the inherent natural variability in forests that results from a variety of natural disturbances. Some of this variability can be introduced into managed landscapes through different silvicultural systems. In the end, however, the choice will depend upon the biological, social, and economic objectives for the stand and the landscape (McComb et al. 1994). The basic premise is that when an ecosystem is managed within its historical range of natural variability, it will remain diverse, resilient, productive, and healthy (Swanson et al. 1994).

Kimmins (1995) defined ecosystem integrity on the basis that structure, species composition, process rates, and the ability to resist change induced by disturbance or stress are all within the characteristic range exhibited by the ecosystem. Management for ecosystem integrity is an essential component of sustainable forestry programs, and clearly forest resource managers will need to consider natural disturbances and their effects on stands and landscapes when formulating management prescriptions. The managers of protected areas especially need to consider the historical and probable future role of natural disturbances, especially regarding their scale and effects and the context (placement of the protected area within the regional matrix).

Research Needs

Apparently the preservation of species depends upon the preservation of natural processes, of which disturbances are an important component (White 1979). Yet we do not fully understand the role of natural disturbances in many locales, and environmental conditions have not necessarily been adequately investigated and documented. We cannot predict the response of all species to natural disturbances because species responses are probably individualistic and influenced by stochastic factors (White 1987).

Because disturbance processes and regimes have varied with historical climate changes and at different spatial and temporal scales, it is not always a simple matter to describe the 'typical' natural disturbance regime currently operating in any location. Indeed, it may be next to impossible to define what is natural at all, given that fully natural ecosystems or environments probably do not exist (Shrader-Frechette and McCoy 1995). The nature of both past and future climate change is uncertain, as are the impacts of changing disturbance regimes (Lertzman et al. 1994; Bergeron and Flannigan 1995).

In spite of the inherent difficulties, some ecological and others philosophical, inherent in natural disturbance research, it is possible to identify a number of areas that require special attention and/or research emphasis:

The relationships between landforms, landscapes, and natural disturbances need to be examined in more detail. Landforms, taken as topography plus surficial material, are the underlying key to the pattern of associated climate, soil, and vegetation. The basic terrain mosaics are modified by a variety of natural disturbances that, in conjunction with secondary succession, serve to distort and sometimes conceal the primary pattern (Rowe 1996). It is recognized that landforms may influence the frequency and pattern of natural disturbance agents such as wildfire, wind, snow, and herbivory, and yet most evidence is only anecdotal or at best qualitative (Swanson et al. 1988).

Our interpretation of what is 'natural' and what is 'harmful' needs to be better understood so that we can deal appropriately with disturbances. Our response to natural disturbances is coloured by our reaction to the harmful environmental consequences of unnatural disturbances such as settlement and conversion of land to agricultural or industrial uses (White 1979). Many ideas about the unnaturalness of disturbance result from our observation of the present altered landscape and from the fact that some natural disturbances operate on a cycle much longer than the human lifetime. Because we rarely see events such as the large wildfires that affected Yellowstone National Park in 1988, they are considered destructive even though they are simply an infrequent process that is integral to the existence of those forest ecosystems (Kimmins 1992).

Pathogens need to be examined beyond the perspective of forest damage. Pathogenic agents of disturbance have received little attention at the landscape scale, especially compared with the more obvious wildfires and blowdowns. With the increasing emphasis on ecosystem management and the maintenance of natural disturbance regimes, the role of pathogens deserves careful consideration (van der Kamp 1991; Castello et al. 1995). Pathogens interact with abiotic disturbance agents to affect the direction and rate of succession, but this too has received little attention.

The effects of our long-term interference in some ecosystems with infrequent natural disturbance regimes is largely unknown and needs to be studied. Interference with natural disturbance processes has sometimes had adverse effects. For example, the suppression of wildfires in interior Douglas-fir and ponderosa pine ecosystems has resulted in increased stand densities, stagnant tree growth, insect and disease outbreaks, increased fuel loadings and wildfire hazard, a decline in understorey herb and shrub cover, and a shift in tree species composition towards the later seral species (Agee 1994). Because the natural disturbance regime of these dry interior ecosystems has been greatly affected by human actions, the adverse impacts and potential consequences are quite evident. In other ecosystems with less frequent natural disturbance regimes, our interference with longer-term natural processes may also have adverse consequences that are, at this point, largely unknown. Wherever ecosystem restoration is clearly required, it will first be necessary to understand the historical range of conditions and natural disturbance regimes that were present before Euro-American settlement (Covington et al. 1994).

There is a need to better understand what occurs in each 'natural' ecosystem so that management practices do not increase the likelihood of epidemics. Epidemics of forest insects and diseases have always occurred, but it is likely that management practices and the spread of exotic organisms have increased the frequency, size, and intensity of many outbreaks. This has resulted from planting tree species not ecologically appropriate for the site, harvesting after the historical rotation ages of certain tree species, establishing susceptible stands and monocultures, not removing diseased overstorey trees, and suppressing fire (Haack and Byler 1993). Given the known synergistic or antagonistic interactions between abiotic and biotic disturbances, more questions can probably be raised than answered about these issues.

More research is needed on approaches to post-disturbance resource management activities. We often carry out such activities to improve the condition of affected ecosystems. In the case of wildfire rehabilitation, this has usually meant timber salvage, felling dead trees, artificial reforestation, and possibly the seeding of grasses and herbs on fireguards and erodible soils. If we intend to emulate natural disturbances in our forest management operations, we should not treat all natural disturbances alike. Some areas should be left to recover naturally while others may require rehabilitation, depending on the specific situation and context. When wildfires are treated, the end result should be a mix of salvaged areas (where feasible), remnant patches of dead trees, remnant patches of live trees (perhaps in riparian zones), areas planted to the appropriate tree species, areas left to revegetate naturally, and areas seeded to native grasses and herbs. Research into this approach may be desirable, but the question is primarily one of changing procedures.

The effects of recent disturbances need to be studied and future changes monitored. One of the principles espoused by Kaufmann et al. (1994) for ecosystem management is that ecosystem processes such as disturbance, succession, evolution, natural extinction, recolonization, fluxes of materials, and other stochastic, deterministic, and chaotic events that characterize the variability in natural ecosystems should be present and functioning. Disturbances, both natural and human-caused, potentially play a large part in the ecology of many of our forests and rangelands. Given the inherent difficulties in determining the detailed characteristics of historical disturbance regimes, the effects of recent disturbances need to be studied and future changes monitored as suggested by Noss (1990), to provide feedback to resource managers.

Literature Cited

Agee, J.K. 1994. Fire and weather disturbances in terrestrial ecosystems of the eastern Cascades. USDA Forest Service General Technical Report PNW-GTR-320. Portland, Ore. 52 pp.

Anderson, L., C.E. Carlson, and R.H. Wakimoto. 1987. Forest fire frequency and western spruce budworm outbreaks in western Montana. Forest Ecology and Management 22:251-260.

Andrews, Major L.A. 1921. The aeroplane's service to forestry. The Illustrated Canadian Forestry Magazine 27(8):419-422, 451-452, 461-462.

Anonymous. 1921. Air reconnaissance of damaged timber limits. Western Lumberman 18(5):30.

Attiwill, P.M. 1994. The disturbance of forest ecosystems: the ecological basis for conservative management. Forest Ecology and Management 63:247-300.

Averill, R.D., L. Larson, J. Saveland, P. Wargo, and J. Williams. 1995. Disturbance processes and ecosystem management. USDA Forest Service, Washington, D.C. 16 pp. [http://www.fs.fed.us/research/disturb.html]

Baker, W.L. 1989. Landscape ecology and nature reserve design in the Boundary Waters Canoe Area, Minnesota. Ecology 70(1):23-35.

–. 1992. The landscape ecology of large disturbances in the design and management of nature reserves. Landscape Ecology 7(3):181-194.

Baker, W.L., and T.T. Veblen. 1990. Spruce beetles and fires in the nineteenth century subalpine forests of western Colorado. Arctic and Alpine Research 22:65-80.

Banner, A., D.V. Meidinger, E.C. Lea, R.E. Maxwell, and B.C. von Sacken. 1996. Ecosystem mapping methods for British Columbia. Pages 97-117 in Richard A. Sims, Ian G.W. Corns, and Karel Klinka (eds.). Global to Local: Ecological Land Classification. 14-17 August 1994, Thunder Bay, Ont. Kluwer Academic Publishers, Dordrecht, Netherlands. Also published in Environmental Monitoring and Assessment 39:97-117.

B.C. Ministry of Forests. 1984. Forest and range resource analysis, 1984. Strategic Studies Branch, Ministry of Forests, Victoria, B.C. Queen's Printer, Victoria, B.C. Page E6. Unpaginated.

B.C. Ministry of Forests, B.C. Ministry of Environment, Lands and Parks, Federal Department of Fisheries and Oceans, and Council of Forest Industries. 1993. British Columbia Coastal Fisheries/Forestry Guidelines. Rev. 3rd ed. Research Branch, Ministry of Forests, Victoria, B.C. 133 pp.

Bell, S. 1994. Visual landscape design training manual. B.C. Ministry of Forests and the British Forestry Commission, Canada-British Columbia Partnership Agreement on Forest Resource Development: FRDA II. Recreation Branch, Ministry of Forests, Victoria, B.C. 166 pp.

Bergeron, Y., and M.D. Flannigan. 1995. Predicting the effects of climate change on fire frequency in the southeastern Canadian boreal forest. Water, Air and Soil Pollution 82:437-444.

Booth, D.L., D.W.K. Boulter, D.J. Neave, A.A. Rotherham, and D.A. Welsh. 1993. Natural forest landscape management: a strategy for Canada. The Forestry Chronicle 69(2):141-145.

Bormann, F.H., and G.E. Likens. 1979a. Catastrophic disturbance and the steady state in northern hardwood forests. American Scientist 67:660-669.

–. 1979b. Pattern and process in a forested ecosystem – disturbance, development and the steady state based on the Hubbard Brook Ecosystem Study. Springer-Verlag, New York. 264 pp.

Bunnell, F.L. 1995. Forest-dwelling vertebrate faunas and natural fire regimes in British Columbia: patterns and implications for conservation. Conservation Biology 9(3):636-644.

Burton, P.J., A.C. Balisky, L.P. Coward, S.G. Cumming, and D.D. Kneeshaw. 1992. The value of managing for biodiversity. The Forestry Chronicle 68(2):225-237.

Cameron, I.T. 1972. Planning guidelines for coast logging operations. Attachment to a letter to licencees of timber sales, tree farm licences, and timber sale harvesting licences in the coast region of British Columbia, dated 29 September 1972. Office of the Chief Forester, Victoria, B.C. File 04155. 5 pp.

Carey, A.B. 1989. Wildlife associated with old-growth forests in the Pacific Northwest. Natural Areas Journal 9:151-162.

Castello, J.D., D.J. Leopold, and P.J. Smallidge. 1995. Pathogens, patterns, and processes in forest ecosystems. BioScience 45(1):16-24.

Cissel, J.H., F.J. Swanson, W.A. McKee, and A.L. Burditt. 1994. Using the past to plan the future in the Pacific Northwest. Journal of Forestry 92(8):30-31,46.

Covington, W.W., R.L. Everett, R. Steele, L.I. Irwin, T.A. Daer, and A.N.D. Auclair. 1994. Historical and anticipated changes in forest ecosystems of the Inland West of the United States. Journal of Sustainable Forestry 2(1/2):13-63.

DeGraaf, R.M., and W.M. Healy. 1993. The myth of nature's constancy – preservation, protection and ecosystem management. Pages 17-28 in Transactions of the Fifty-eighth North American Wildlife and Natural Resources Conference. 19-24 March 1993, Washington, D.C. Wildlife Management Institute, Washington, D.C.

Diaz, N., and D. Apostol. 1992. Forest landscape analysis and design. USDA Forest Service Pacific Northwest Region Publication R6 ECO-TP-043-92. Unpaginated.

Diaz, N., and Fellin, D.G. 1980. A review of some interactions between harvesting, residue management, fire, and forest insects and diseases. Pages 335-414 in Environmental consequences of timber harvesting in rocky mountain coniferous forests. Symposium proceedings. 11-13 September 1979, Missoula, Mont. USDA Forest Service General Technical Report INT-90. Ogden, Utah. 526 pp.

Diaz, N.M., and S. Bell. 1997. Landscape analysis and design. Pages 255-269 in K.A. Kohm and J.F. Franklin (eds.). Creating a forestry for the 21st century: the science of ecosytem management. Island Press, Washington, D.C. 491 pp.

Flannigan, M.D., T.J. Lynham, and P.C. Ward. 1989. An extensive blowdown occurrence in northwestern Ontario. Pages 65-71 in Proceedings: 10th Conference on Fire and Forest Meteorology, 17-21 April 1989, Ottawa, Ont. Forestry Canada and Environment Canada, n.p.

Foote, M.J. 1983. Classification, description, and dynamics of plant communities after fire in the taiga of interior Alaska. USDA Forest Service Research Paper PNW-307. Portland, Ore. 108 pp.

Forman, R.T.T. 1983. An ecology of the landscape. BioScience 33(9):535.

Forman, R.T.T., and M. Godron. 1981. Patches and structural components for a landscape ecology. BioScience 31(10):733-740.

Franklin, J.F. 1992. Forest stewardship in an ecological age. The ninth annual C.E. Farnsworth Memorial Lecture, Faculty of Forestry, State University of New York College of Environmental Science and Forestry, Syracuse, N.Y. 26 pp.

–. 1993. Preserving biodiversity: species, ecosystems, or landscapes? Ecological Applications 3(2):202-205.

Franklin J.F., and R.T.T. Forman. 1987. Creating landscape patterns by forest cutting: ecological consequences and principles. Landscape Ecology 1(1):5-18.

Franklin, J.F., H.H. Shugart, and M.E. Harmon. 1987. Tree death as an ecological process. BioScience 37(8):550-556.

Furniss, R.L., and V.M. Carolin. 1980. Western forest insects. USDA Forest Service Miscellaneous Publication No. 1339. Washington, D.C. 661 pp.

Galindo-Leal, C., and F. Bunnell. 1995. Ecosystem management: implications and opportunities of a new paradigm. The Forestry Chronicle 71(5):601-606.

Geiszler, D.R., R.I. Gara, C.H. Driver, V.F. Gallucci, and R.E. Martin. 1980. Fire, fungi, and beetle influences on a lodgepole pine ecosystem of south-central Oregon. Oecologia 46:239-243.

Gillis, A.M. 1990. The new forestry – an ecosystem approach to land management. BioScience 40(8):558-562.

Grumbine, R.E. 1994. What is ecosystem management? Conservation Biology 8(1):27-38.

Guie, H.D. 1921. Washington's forest catastrophe. American Forestry 27(330):379-382.

Haack, R.A., and J.W. Byler. 1993. Insects & pathogens – regulators of forest ecosystems. Journal of Forestry 91(9):32-37.

Hann, W.J. 1992. Management for landscape and ecosystem biodiversity. Pages 63-71 in Angela G. Evenden (comp.). Proceedings – Northern Region Biodiversity Workshop.

11-13 September 1990, Missoula, Mont. Northern Region, USDA Forest Service, Missoula, Mont.

Hansen, A.J., T.A. Spies, F.J. Swanson, and J.L. Ohmann. 1991. Conserving biodiversity in managed forests – lessons from natural forests. BioScience 41(6):382-392.

Harris, L.D., and C. Maser. 1984. Animal community characteristics. Pages 44-68 in The fragmented forest: island biogeography theory and the preservation of biotic diversity. University of Chicago Press, Chicago, Ill. 211 pp.

Hodgins, H.J. (compiler). 1964. Forest resources. Pages 332-333 in D.B. Turner (ed.). Inventory of the natural resources of British Columbia. The British Columbia Natural Resources Conference, n.p. Tables 17 and 18. 610 pp.

Johnson, D.R., and J.K. Agee. 1988. Introduction to ecosystem management. Pages 3-14 in J.K. Agee and D.R. Johnson (eds.). Ecosystem management for parks and wilderness. University of Washington Press, Seattle, Wash.

Johnson, E.A., G.I. Fryer, and M.J. Heathcott. 1990. The influence of man and climate on frequency of fire in the interior wet belt forest, British Columbia. Journal of Ecology 78(2):403-412.

Kaufmann, M.R., R.T. Graham, D.A. Boyce Jr., W.H. Moir, L. Perry, R.T. Reynolds, R.L. Bassett, P. Mehlhop, C.B. Edminster, W.H. Block, and P.S. Corn. 1994. An ecological basis for ecosystem management. USDA Forest Service General Technical Report RM-246. Fort Collins, Colo. 22 pp.

Kelsall, J.P., E.S. Telfer, and T.D. Wright. 1977. The effects of fire on the ecology of the boreal forest, with particular reference to the Canadian north: a review and selected bibliography. Occasional Paper No. 32, Canadian Wildlife Service, Fisheries and Environment Canada, Ottawa, Ont. 58 pp.

Kimmins, H. 1992. Balancing act: environmental issues in forestry. UBC Press, Vancouver, B.C. 244 pp.

Kimmins, J.P. 1995. Sustainable development in Canadian forestry in the face of changing paradigms. The Forestry Chronicle 71(1):33-40.

Lertzman, K.P., and C.J. Krebs. 1991. Gap-phase structure of a subalpine old-growth forest. Canadian Journal of Forest Research 21:1730-1741.

Lertzman, K., V. Kurtz, T.A. Spies, and F.J. Swanson. 1994. Ecosystems dynamics. Page 6 in Conference on the Environment and People of the Coastal Temperate Rain Forest. 29-31 August 1994, Whistler, B.C.

Lertzman, K., T. Spies, and F. Swanson. 1997. From ecosystem dynamics to ecosystem management. Pages 261-382 in P.K. Schoonmaker, B. von Hagen, and E.C. Wolf (eds.). The rain forests of home: profile of a North American bioregion. Island Press, Washington, D.C. 455 pp.

MacKinnon, A., D. Meidinger and K. Klinka. 1992. Use of the biogeoclimatic ecosystem classification system in British Columbia. The Forestry Chronicle 68(1):100-120.

McComb, W., J. Tappeiner, L. Kellogg, C. Chambers, and R. Johnson. 1994. Stand management alternatives for multiple resources: integrated management experiments. Pages 71-86 in Mark H. Huff, L.K. Norris, J.B. Nyberg, and N.L. Wilkin (coord.). Expanding horizons of forest management: proceedings of the Third Habitat Futures Workshop. USDA Forest Service General Technical Report PNW-GTR-336. Portland, Ore.

Morgan, P., G.H. Aplet, J.B. Haufler, H.C. Humphries, M.M. Moore, and W.D. Wilson. 1994. Historical range of variability: a useful tool for evaluating ecosystem change. Journal of Sustainable Forestry 2(1/2):87-111.

Mulholland, F.D. 1937. The forest resources of British Columbia. British Columbia Forest Service, Department of Lands. Queen's Printer, Victoria, B.C. 153 pp.

Norris, L.A., H. Cortner, M.R. Cutler, S.G. Haines, J.E. Hubbard, M.A. Kerrick, W.B. Kessler, J.C. Nelson, R. Stone, J.M. Sweeney, and L.W. Hill. 1992. Task force report on sustaining long-term forest health and productivity. Society of American Foresters, Bethesda, Md. 83 pp.

Noss, R.F. 1983. A regional landscape approach to maintain diversity. BioScience 33(11):700-706.

–. 1990. Indicators for monitoring biodiversity: a hierarchical approach. Conservation Biology 4(4):355-364.

Odum, E.P. 1971. Fundamentals of ecology. 3rd ed. W.B. Saunders Company, Philadelphia, Pa. 574 pp. (p. 8)

Oliver, C.D. 1992. A landscape approach – achieving and maintaining biodiversity and economic productivity. Journal of Forestry 90(9):20-25.

–. 1995. Rebuilding biological diversity at the landscape level. Pages 95-115 in Proceedings of the conference – Forest Health and Fire Danger in Inland Western Forests. 8-9 September 1994, Spokane, Wash.

Oliver, C.D., and B.C. Larson. 1990. Forest stand dynamics. McGraw-Hill, New York. 467 pp. (pp. 355-383)

Perry, D.A. 1988. Landscape patterns and forest pests. Northwest Environmental Journal 4:213-228.

Perry, D.A., and M.P. Amaranthus. 1997. Disturbance, recovery and stability. Pages 11-30 in K.A. Kohm and J.F. Franklin (eds.) Creating a forestry for the 21st century: the science of ecosystem management. Island Press, Washington, D.C. 491 pp.

Petraitis, P.S., R.E. Latham, and R.A. Niesenbaum. 1989. The maintenance of species diversity by disturbance. The Quarterly Review of Biology 64(4):393-418.

Pickett, S.T.A., and J.N. Thompson. 1978. Patch dynamics and the design of nature reserves. Biological Conservation 13:27-37.

Pickett, S.T.A., and P.S. White. 1985. Patch dynamics: a synthesis. Pages 371-384 in S.T.A. Pickett and P.S. White (eds.). The ecology of natural disturbance and patch dynamics. Academic Press, Orlando, Fla.

Pojar, J., K. Klinka, and D.V. Meidinger. 1987. Biogeoclimatic ecosystem classification in British Columbia. Forest Ecology and Management 22:119-154.

Pojar, J., N. Diaz, D. Steventon, D. Apostol, and K. Mellen. 1994. Biodiversity planning and forest management at the landscape scale. Pages 55-70 in Mark H. Huff, L.K. Norris, J.B. Nyberg, and N.L. Wilkin (coord.). Expanding horizons of forest management: proceedings of the Third Habitat Futures Workshop. USDA Forest Service General Technical Report PNW-GTR-336. Portland, Ore.

Probst, J.R., and T.R. Crow. 1991. Integrating biological diversity and resource management. Journal of Forestry 89(2):12-17.

Quirk, W.A., and D.J. Sykes. 1971. White spruce stringers in a fire-patterned landscape in interior Alaska. Pages 179-197 in C.W. Slaughter, R.J. Barney, and G.M. Hansen (eds.). Fire in the northern environment – a symposium. Proceedings of a symposium sponsored by Alaska Forest Fire Council and Alaska Section, Society of American Foresters. 13-14 April 1971, College, Alaska. USDA Forest Service, Pacific Northwest Forest and Range Experiment Station, Portland, Ore.

Reice, S.R. 1994. Nonequilibrium determinants of biological community structure. American Scientist 82(5):424-435.

Revised Statutes of British Columbia. 1979. Ministry of Forests Act. RS Chap. 272, 28 Eliz. 2. Queen's Printer, Victoria, B.C. 4 pp.

Romme, W.H., and D.H. Knight. 1982. Landscape diversity: the concept applied to Yellowstone Park. BioScience 32:664-670.

Rowe, J.S. 1992. The ecosystem approach to forestland management. The Forestry Chronicle 68(1):222-224.

–. 1993. Biodiversity at the landscape level. Manuscript prepared for 'Measuring Biodiversity for Forest Policy and Management,' University of British Columbia Centre for Applied Conservation Biology, Vancouver, B.C., 24 February 1994. 15 pp.

–. 1996. Land classification and ecosystem classification. Pages 11-20 in Richard A. Sims, Ian G.W. Corns and Karel Klinka (eds.). Global to Local: Ecological Land Classification. 14-17 August 1994, Thunder Bay, Ont. Kluwer Academic Publishers, Dordrecht, Netherlands. Also published in Environmental Monitoring and Assessment 39:11-20.

Safranyik, L., D.M. Shrimpton, and H.S. Whitney. 1974. Management of lodgepole pine to reduce losses from the mountain pine beetle. Environment Canada Forestry Service, Forestry Technical Report 1. Victoria, B.C. 24 pp. Reprinted 1976.

Sedell, J.R., P.A. Bisson, F.J. Swanson, and S.V. Gregory. 1988. What we know about large trees that fall into streams and rivers. Pages 47-81 in C. Maser, R.F. Tarrant, J.M. Trappe, and J.F. Franklin (eds.). From the forest to the sea: a story of fallen trees. USDA Forest

Service General Technical Report PNW-GTR-229. Published in cooperation with the USDI Bureau of Land Management. Portland, Ore. 153 pp.

Shrader-Frechette, K.S., and E.D. McCoy. 1995. Natural landscapes, natural communities and natural ecosystems. Forest & Conservation History 39(3):138-142.

Smith, J.H.G. 1981. Fire cycles and management alternatives. Pages 511-531 in H.A. Mooney, T.M. Bonnicksen, N.L. Christensen, J.E. Lotan, and W.A. Reiners (tech. coords.). Proceedings of the conference: Fire Regimes and Ecosystem Properties. USDA Forest Service General Technical Report WO-26. Washington, D.C.

Sousa, W.P. 1984. The role of disturbance in natural communities. Annual Review of Ecology and Systematics 15:353-391.

Spandli, I. 1993. Addendum to a memorandum to all branch directors on the subject of harvesting over the last 140 years. From D.E. Gilbert, Director, Inventory Branch, dated 17 August 1993. Ministry of Forests, Victoria, B.C. File 13000. 7 pp.

Spies, T.A., and J.F. Franklin. 1988. Old growth and forest dynamics in the Douglas-fir region of western Oregon and Washington. Natural Areas Journal 8(3):190-201.

Spies, T.A., J.F. Franklin, and T.B. Thomas. 1988. Coarse woody debris in Douglas-fir forests of western Oregon and Washington. Ecology 69:1689-1702.

Spies, T.A., W.J. Ripple, and G.A. Bradshaw. 1994. Dynamics and pattern of a managed coniferous forest landscape in Oregon. Ecological Applications 4(3):555-568.

Sprugel, D.G. 1991. Disturbance, equilibrium, and environmental variability: what is 'natural' vegetation in a changing environment? Biological Conservation 58:1-18.

Stathers, R.J., T.P. Rollerson, and S.J. Mitchell. 1994. Windthrow handbook for British Columbia. Ministry of Forests Research Program Working Paper 9401. Victoria, B.C. 31 pp.

Stuart-Smith, K., and D. Hebert. 1996. Putting sustainable forestry into practice at Alberta-Pacific. Canadian Forest Industries (April/May 1996):57-60.

Swanson, F.J., T.K. Kratz, N. Caine, and R.G. Woodmansee. 1988. Landform effects on ecosystem patterns and processes. BioScience 38(2):92-98.

Swanson, F., and D. Berg. 1991. The ecological roots of new approaches to forestry. Forest Perspectives 1(3):6-8.

Swanson, F.J., and J.F. Franklin. 1992. New forestry principles from ecosystem analysis of Pacific Northwest forests. Ecological Applications 2(3):262-274.

Swanson, F.J., J.A. Jones, D.O. Wallin, and J.H. Cissel. 1994. Natural variability – implications for ecosystem management. Pages 80-94 in M.E. Jensen and P.S. Bourgeron (tech. eds.). Volume II. Ecosystem management: principles and applications. Eastside Forest Ecosystem Health Assessment, Richard L. Everett, Assessment Team Leader. USDA Forest Service General Technical Report PNW-GTR-318. Portland, Ore.

Tappeiner, J.C., D. Lavender, J. Walstad, R.O. Curtis, and D.S. DeBell. 1997. Silvicultural Systems and regeneration methods: current practices and new alternatives. Pages 151-164 in K.A. Kohm and J.F. Franklin (eds.). Creating a forestry for the 21st century: the science of ecosystem management. Island Press, Washington, D.C. 491 pp.

Taylor, A.H. 1990. Disturbance and persistence of sitka spruce (*Picea sitchensis* [Bong.] Carr.) in coastal forests of the Pacific Northwest, North America. Journal of Biogeography 17:47-58.

Thomas, J.W. (tech. ed.). 1979. Wildlife habitats in managed forests: the Blue Mountains of Oregon and Washington. Agriculture Handbook No. 553. USDA Forest Service, Portland, Ore. 512 pp.

Turner, M.G. 1989. Landscape ecology: the effect of pattern on process. Annual Review of Ecology and Systematics 20:171-197.

Turner, M.G., W.H. Romme, and R.H. Gardner. 1994. Landscape disturbance models and the long-term dynamics of natural areas. Natural Areas Journal 14(1):3-11.

Urban, D.L., R.V. O'Neill, and H.H. Shugart Jr. 1987. Landscape ecology. BioScience 37(2):119-127.

van der Kamp, B.J. 1991. Pathogens as agents of diversity in forested landscapes. The Forestry Chronicle 67(4):353-354.

Veblen, T.T., K.S. Hadley, E.M. Nel, T. Kitzberger, M. Reid, and R. Villalba. 1994. Disturbance regime and disturbance interactions in a Rocky Mountain subalpine forest. Journal of Ecology 82:125-135.

Vogl, R.J. 1980. The ecological factors that produce perturbation-dependent ecosystems. Pages 63-94 in John Cairns Jr. (ed.). The recovery process in damaged ecosystems. Ann Arbor Science Publishers/The Butterworth Group, Ann Arbor, Mich. 167 pp.

Watt, A.S. 1947. Pattern and process in the plant community. Journal of Ecology 35(1&2): 1-22.

Weaver, H. 1974. Effects of fire on temperate forests: western United States. Pages 279-319 in T.T. Kozlowski and C.E. Ahlgren (eds.). Fire and ecosystems. Academic Press, New York.

Weetman, G.F. 1957. The chief causes of blowdown in pulpwood stands in eastern Canada. Woodlands Information Report No. 1, Pulp and Paper Research Institute of Canada, Montreal, Que. 8 pp.

Wiedenmann, K.R. 1992. Shasta Costa: from a new perspective. Pages 25-30 in Angela G. Evenden (comp.). Proceedings – Northern Region Biodiversity Workshop. 11-13 September 1990, Missoula, Mont. Northern Region, USDA Forest Service, Missoula, Mont.

White, P.S. 1979. Pattern, process and natural disturbance in vegetation. The Botanical Review 45(3):229-299.

–. 1987. Natural disturbance, patch dynamics, and landscape pattern in natural areas. Natural Areas Journal 7(1):14-22.

White, P.S., and S.P. Bratton. 1980. After preservation: philosophical and practical problems of change. Biological Conservation 18:241-255.

White, P.S., and S.T.A. Pickett. 1985. Natural disturbance and patch dynamics: an introduction. Pages 3-13 in S.T.A. Pickett and P.S. White (eds.). The ecology of natural disturbance and patch dynamics. Academic Press, Orlando, Fla.

Wright, H.A., and A.W. Bailey. 1982. Fire ecology – United States and southern Canada. John Wiley and Sons, N. Y. 499 pp.

Wright, K.H., and G.M. Harvey. 1967. The deterioration of beetle-killed Douglas-fir in western Oregon and Washington. USDA Forest Service Research Paper PNW-50. Portland, Ore. 20 pp.

2
Spatial Patterns in Forested Landscapes: Implications for Biology and Forestry
Marvin Eng

Definitions

The study of spatial patterns in forested landscapes is largely included in the field of landscape ecology. Landscape ecology is relatively new to North Americans. The first issue of the journal *Landscape Ecology* (published by the International Association of Landscape Ecology) is dated 1988, and the essential text on the subject is Forman and Godron (1986). The 'language' of the field suffers from the common problems of new sciences.

First, many terms are simply nothing but pedantic jargon. For example, an isodiametrical patch (Zipperer 1993) is simply a circular patch. Second, the definitions of terms are often context-dependent and open to interpretation. For example the term *matrix* is defined as 'the most extensive and contiguous landscape element that exerts the most control over landscape function' (Forman and Godron 1986). The matrix could therefore change over time, as the landscape changes. In addition, one cannot determine a priori what the matrix will be. In an agricultural landscape with forested patches (such as parts of the south Okanagan), the matrix is the agricultural cover; in a forested landscape with agricultural patches (such as parts of the Peace River area), the matrix is the forest.

Finally, the field is replete with 'concept clusters' (Peters 1991). A concept cluster is a term that has a group of similar, but not identical, definitions. In landscape ecology, the most obvious of these is the term *landscape* itself. This word has a somewhat different meaning for almost everyone who uses it, principally because it is often defined using terms that are also usually poorly defined.

In this section an attempt is made to identify the single, most generic (context-insensitive) definition for some of the more important terms and concept clusters used in the landscape ecology literature. The rather large body of pure jargon used in the field is not dealt with here.

Extent: with reference to scale, the size of the study area or the duration of time under consideration (Turner and Gardner 1991).

Fragmentation: the process of creating an increasingly complex mosaic of patches as a result of disturbances, including human activity (Li et al. 1993).

Grain: with reference to scale, the finest level of spatial or temporal resolution within a given data set (Turner and Gardner 1991).

Landscape: a complex of systems, formed by the activity of rock, air, water, plants, animals, and man, that by its physiognomy forms a recognizable entity (Zonneveld 1979).

Landscape ecology: a branch of ecology that attempts to develop an understanding of the development and dynamics of pattern in ecological phenomena, the role of disturbances in ecosystems, and characteristic spatial and temporal scales of ecological events (Urban et al. 1987).

Matrix: the most extensive and contiguous landscape element that exerts the most control over landscape function (Forman and Godron 1986).

Natural variability: the composition, structure, and dynamics of ecosystems before the influence of European settlers. The range of natural variability is characterized by the range of ecosystem conditions and the variety of seral classes that result from the regime that affects the area (frequency, severity, spatial arrangement) (Swanson et al. 1993).

Patch: a landscape element that differs from the matrix with respect to composition and/or condition (Forman and Godron 1986).

Scale: the spatial or temporal dimensions of an object or process. Scale is characterized by both grain and extent (Turner et al. 1991).

An example of patches (clearcuts) showing how they differ from the surrounding matrix (forest). (Photograph by Scott Harrison)

Of these definitions, the most problematic are those for landscape, matrix, and patch. The key to the definition of landscape provided above is the word *recognizable*. Much of recent efforts in landscape ecology is directed towards the recognition of landscapes, particularly at different scales (Turner et al. 1991). The term, like its progenitor, *ecosystem*, is entirely scale- and context-dependent. A landscape for a bark beetle is entirely different from a landscape for a wolf. In British Columbia, a landscape (unit) for a forest manager is more or less legally defined as a watershed, or series of contiguous watersheds, of 5,000-100,000 ha in size with a relatively homogeneous natural disturbance regime (B.C. Ministry of Forests 1996).

The original distinction between *matrix* and *patch* (as defined above) appears to have been of value in structuring early thoughts in North American landscape ecology (e.g., Franklin and Forman 1987). However, advances in theory and mapping technology have resulted in the blurring of this distinction. Originally, the term *matrix* was applied to more or less homogeneous tracts of agricultural land or mature forest. The corresponding patches were woodland remnants or clearcuts, respectively. This was done to simplify the process of landscape analysis. We now have the technological capability to treat the matrix as a series of patches that are similar in some respects but different in others. The matrix can now be viewed, at most, as the single largest patch in the landscape. A patch is more or less definable as the smallest resolvable element in the landscape. Operationally, this would be a polygon on the map of the landscape.

Background
The pattern of forest harvesting in British Columbia has changed substantially over time. In the distant past, Aboriginal peoples selectively harvested individual trees for constructing dwellings and boats. This pattern was replicated by early non-natives who needed timber for building forts and homes and for repairing and building ships. As non-native populations grew, forest cover was seen as an impediment to agricultural development, particularly in the lower reaches of the Fraser River and on Vancouver Island. This led to the clearing of large areas of forest and their subsequent conversion to agricultural uses. Only in this century has there been sufficient demand for wood fibre to result in large-scale forest harvesting (Robson 1995).

Initially, the landscape patterns created by forest harvesting were driven largely by economic considerations. The common harvesting pattern, particularly in coastal environments, was known as the 'progressive clearcut.' Forest harvesting began at the lower end of a valley system and progressed up the bottom of the valley, with harvesting of the high-value timber associated with the alluvial and toe slope sites. Harvesting of these areas resulted in the creation of a road network that then enabled economical harvesting of forests further up the slope. Depending on the rate at which harvesting progressed, this method resulted in a forest pattern consisting of large areas of more or less similar-aged second growth in the lower areas of the valley surrounded by 'original' forest on the upper slopes that could not be harvested economically.

In the Pacific Northwest (Washington, Oregon, and California), and subsequently in British Columbia, the practice of progressive clearcutting was criticized as the area of land subjected to this practice increased. The criticisms pointed to the effects of progressive clearcuts on hydrologic regimes, sediment production, and wildlife habitat, and to the visual impact of the practice. In response, managers of public forest land adopted a forest-harvesting pattern consisting of small (10-20 ha) cutblocks dispersed throughout a management area in space and time. This forest-harvesting pattern was seen to have the following advantages over progressive clearcuts (Wallin et al. 1994):

- promotion of regeneration through seed rain, thus reducing the need for planting
- development of an extensive road network, enabling effective fire suppression in the harvested and unharvested forest
- creation of edge habitat and small patches of early seral stage habitat favoured by many species of game
- dispersal and reduction of the effects of forest harvesting on hydrologic regimes and sediment production
- reduction of the visual impact of forest harvesting.

The forest-harvesting pattern of dispersed cutblocks has been promoted in British Columbia by rules that place constraints on the age of forest in adjacent harvest blocks and by the Forest Practices Code of British Columbia Act (Province of British Columbia 1995a). This act limits cutblock sizes to 40 ha in coastal B.C. and 60 ha in the interior. There are also 'adjacency' rules to ensure that cutblocks will be dispersed in space and time. However, the implementation of this pattern of dispersed cutblocks is being critically examined because many of the original objectives of the practice have been met or superseded. For example, the practice of regeneration through seed rain has been superseded by the practice of planting genetically, ecologically, and economically suitable tree seedlings (Wallin et al. 1994). Also, it is recognized that although roads enable fire protection, they are often the cause of many ignitions because of increased public access. Finally, the objective of improving habitat for edge species and for 'game' species associated with early seral stages has been superseded by the objective of maintaining all native plants and animals over their historical ranges (B.C. Ministry of Environment, Lands and Parks, 1994).

There is an ecological concern that dispersed cutblocks will cause forest fragmentation (Harris 1984; Franklin and Forman 1987). This would potentially have a negative impact on species that are dependent on forest interior conditions. The fragmentation of forests into small patches will reduce the amount of interior habitat, and the increase in edges will cause changes in forest structure and composition through competition and changes in microclimate (Saunders et al. 1991; Chen et al. 1993). (For more information on interior habitat and edges, refer to Chapters 5 and 8.)

In the past, the landscape patterns created by forest management occurred without conscious planning. They were simply the sum of activities at the stand scale (individual cutblocks). Two schools of thought now exist with regard to landscape planning. The first school describes future landscape patterns in terms of certain desired products (e.g., wood fibre, habitat) and known ecosystem processes. The second school bases its future patterns on historical patterns. 'This point of view reflects the fact that we cannot even name all of the species in the landscape, much less rationally plan for their habitat needs and ecosystem functions. A premise of this approach is that native species have adapted to the disturbance events and resulting range of habitat patterns of the past thousands of years' (Cissel et al. 1994). The fundamental problem with the first approach is that it presents far too optimistic a view of our abilities as scientists and 'ecosystem engineers.' However, the second approach also leaves us casting about for a method of setting landscape-level management objectives.

Ecological Principles

Scale and Ecological Hierarchy

Understanding scale is central to understanding in ecology (Levin 1992) and to the development of sustainable practices of natural resource use (Ludwig et al. 1993; Lee 1994). Ludwig et al. (1993) focus on mismatches in temporal scale that result in unsustainable exploitation of resources. In the forestry context, the problem is that 'the discounted present value is still the backbone of most significant forest management planning models' (Holgrem 1995). Therefore, harvesting at an unsustainable rate can be rational from an economic point of view if the earnings from harvesting appreciate in value more rapidly than the forest would regenerate. Lee (1994) generalizes the argument to include mismatches between the responsibility of the resource exploiter and spatial and functional scales. In the forestry context, if the spatial or temporal scale of forest harvesting does not match the spatial or temporal scale of natural processes in the forest, the people exploiting the resource may not be the people who pay the costs of the exploitation. Generally in such situations we cannot expect these forests to be ecologically sustainable.

'The problem of pattern and scale is the central problem in ecology, unifying population biology and ecosystems science, and marrying basic and applied ecology' (Levin 1992). The fundamental difficulty is that each species observes its environment on its own unique set of scales of space and time. Various life history traits (dispersal, dormancy, timing of reproduction) have the effect of modifying the scale at which the organism perceives the environment. Differential response to variability results in a partitioning of resources that enhances the coexistence of species. However, the scales we study and manage are often imposed by our perceptual abilities or technologies rather than what is appropriate to the question at hand. Environmental variability has meaning only relative to the scale of observation.

'Many ecologists focus their questions on small scale questions amenable to experimental tests and remain oblivious to the larger scale processes which may account for the patterns they study' (Levin 1992). In models of global climate, however, large-scale processes are overemphasized at the expense of the fine scale (Schneider 1989). Linkages among scales are more common in marine ecology, but few terrestrial examples are available (Schimmel et al. 1990).

No single scale is appropriate for study; systems show variability on a range of spatial, temporal, and organizational scales. The key to prediction is explanation of observed patterns (otherwise one must evaluate each new stress on each system *de novo*), but the processes that create the patterns often operate at different scales from the observed patterns (Turner 1989; O'Neill et al. 1991; Levin 1992).

To predict at different scales, 'we must learn how to aggregate and simplify, retaining essential information, without getting bogged down in unnecessary detail' (Levin 1992). Rather than try to choose a single 'correct' scale, however, we must try to understand how things change across scales. Just because there is no single correct scale does not mean any scale will do. This explains the recent fascination in ecology with fractals, which emphasize scale-dependence of data and phenomena and point towards potential theories of scaling (Mandelbrot 1983; Sugihara and May 1990). For the more mathematically inclined, there are a variety of spatial statistics that attempt to describe how patterns change across scales. It is beyond the scope of this chapter to review these statistics, but the more important ones are as follows:

- blocking techniques (Greig-Smith 1964; Hill 1973; Ludwig 1979)
- semivariograms and correlograms (Sokal et al. 1989)
- spectral analysis (Chatfield 1984; Davis 1986)
- moving window analysis (Delcourt and Delcourt 1988)
- allometry (Brown and Nicoletti 1991; Harvey and Pagel 1991).

These scale concepts are as true for planning forestry operations as they are for biology. The Scientific Panel for Sustainable Forest Practices in Clayoquot Sound (1995) has indicated that 'planning at a variety of spatial and temporal scales is critical at all stages of forest ecosystem management.' Predicting and planning at a variety of scales requires first a conceptual framework for dividing the world into discrete scales. Ecological hierarchy theory provides the basis for this framework.

Urban et al. (1987) review the concepts of hierarchy theory as applied to landscape ecology: 'Hierarchically organized systems can be divided, or decomposed, into discrete functional components operating at different scales.' The components at any given level interact with each other to generate higher-level behaviours; these higher levels control the behaviours at lower levels. 'A component at a given level of a hierarchy experiences as a variable only those patterns that are similarly scaled in rate, as well as in size. Lower level dynamics are so fast that they are experienced as averages, higher level dynamics are too slow to be experienced as variable.'

Of course, natural systems are not perfectly decomposable into discrete levels. However, O'Neill et al. (1991) believe that hierarchical organization in nature should be a natural consequence of the nonlinear rates of interaction among biotic and abiotic components. They hypothesize that interacting components (at the same hierarchical level) will operate at similar dynamic rates and will be relatively isolated from lower levels and constrained by higher levels. Dynamics will therefore be grouped into distinct levels or scales rather than be uniformly distributed between the fastest and slowest rates. As a result the hierarchy of patterns should reflect the hierarchy of process rates.

The components of the hierarchy are organized into levels according to functional scale. Events that occur at a certain level have a characteristic frequency and spatial scale. In general, the higher levels are usually larger units and operate more slowly than the lower levels. The higher levels also tend to constrain the lower levels.

Urban et al. (1987) give an example of a proposed hierarchical structure for an eastern North American deciduous forest. In their example, the tree gap is the lowest level, with a spatial scale equalling the size of a tree and a temporal scale equalling the life span of a tree. The next highest level is the stand, which is a mosaic of gap-sized patches, often bounded by environmental constraints (such as a mesic cove) or a larger disturbance (such as fire). Urban et al. (1987) propose that a watershed should be the next level, because 'stands within a watershed share a similar resource base and interact more among themselves than they do with stands in other watersheds.' The final level, a landscape, would be a group of interacting watersheds bounded by physiographic features (such as mountain ranges governing climate). This can be summarized as follows:

Level	Boundary definition	Scale
Landscape	Physiographic provinces Changes in land use or disturbance regimes	10,000 ha
Watershed	Local drainage basins Topographic divides	100 to 1,000 ha
Stand	Topographic positions Disturbance patches	1 to 10 ha
Gap	Large-tree influences	0.01 to 0.1 ha

This is a rate-structured hierarchy: components within one level interact more with each other than with components of other levels. This rule defines horizontal and vertical structure. Interactions at one level generate behaviours of the components at the next level. Note that the names of the 'levels' are those of Urban et al. (1987). Other authors use different ways to define the same words.

The hierarchy, as described, is also perfectly nested in terms of spatial scale (but not necessarily in terms of function): each patch is an integral whole and part of the next highest level. Levels in a nested hierarchy are defined by ordering criteria: measured attributes (tree species biomass) and constraints (sunlight and nutrients). Constraint and interaction may be mutually reinforcing as ordering criteria (for example, stands defined on topographic moisture may be joined by seed dispersal into landscapes).

Urban et al. (1987) point out that the hierarchies are largely a human construct, and they 'are constructed in relation to the specified phenomenon of interest.' For example, if we are more interested in birds than in trees, territories might be the smallest unit rather than treefall gaps.

Hierarchy theory should provide an understanding of the processes that operate in the system, but the processes must be considered in context. For example, the process of gap formation generates available light in the gap and may allow prediction of the shade tolerance of species that will dominate. However, other factors determine which shade-tolerant species will dominate because there are many other constraints (nutrients, water, disturbance, seed sources) on individual tree growth besides available light. The factors producing these constraints are usually spatially scaled at a higher level and therefore provide context for the behaviour of the lower level. The more constraints we consider, the better our predictive ability. 'The reference level is the scale on which the phenomenon is witnessed as an interesting event. Once specified, the event has its mechanistic explanation at the next lowest level, and its significance in the context of higher-level constraints' (Urban et al. 1987).

Although it is relatively easy to imagine that we know what the levels in the hierarchy are, as in the example of Urban et al. (1987), there is no

(a)

(b)

Examples of a (a) landscape,
(b) stand, (c) watershed, and
(d) gap. (Photographs by Les
Peterson [a], Joan Voller [b,c],
and John Parminter [d])

(c)

(d)

reason to expect that we can simply 'perceive' a rate-structured hierarchy in nature. O'Neill et al. (1991) provide an example of how some of the spatial statistics mentioned previously are used to attempt to locate a hierarchical pattern in the distribution of plants at three different sites in the western United States. They calculated four spatial statistics on transect data of vegetation cover. Their results were simultaneously encouraging and disheartening. They were able to detect multiple scales of vegetation pattern at all three sites. They noted that field data over 4 orders of magnitude (decimetre measurements over kilometres) were required to detect scales over 2 orders of magnitude (10-1,000 m), and that any single method of data analysis always missed some of the patterns present at each site.

They concluded that detection of multiple scales of pattern was only the first step in demonstrating hierarchical organization, because abiotic factors alone (climate, topography, microsite) could result in the observed patterns. The method did not reveal processes but encouraged speculation and hypothesis. For example, higher-level constraints were hypothesized to have imposed patterns that occurred across an entire community (apparently at 300-700 m). Patterns shared by two or more species but not by all species might indicate competition. Phase differences caused by disturbances should be seen at the same scale across all species. 'Scale patterns can be caused by patterns in soils, competitive interactions, unique adaptations to local conditions, and disturbance-recovery processes.' Each process would have a 'signature' of pattern and species involved.

Modern forest management has impacts that occur at all scales from molecular to global. In the context of the spatial patterns created by forestry, the challenge for forest managers and researchers is to determine which scales are relevant to a particular forestry activity or research endeavour. We should not be constrained by preconceived notions about what scales (levels of organization) exist in nature. The fields of spatial statistics and hierarchy theory appear to have some potential in aiding the identification of scales.

Fragmentation
Habitat fragmentation has been proposed as the most serious threat to biological diversity confronting humankind today (Wilcox and Murphy 1985; Lovejoy et al. 1986; Wilcove et al. 1986). The case can easily be made that direct habitat loss is a much more serious issue, and that by focusing on the problem of habitat fragmentation we are giving in to the need to apply remedial measures in the face of the overwhelming alteration of the environment by humans. However, given that we, as a society, are going to alter substantial portions of our forested landscapes through forest harvesting, we should probably be concerned about minimizing the effect of fragmentation on the remaining forest.

Fragmentation produces remnant vegetation patches surrounded by a matrix of different vegetation. These remnant patches may be isolated by varying degrees. (Photograph by John Parminter)

The concern over fragmentation represents a reversal for wildlife managers and biologists, who traditionally promoted the creation of 'edge habitat' (primarily for game species) (Leopold 1933; Yoakum and Dasmann 1971) that is characteristic of at least the early stages of fragmentation. More recently, wildlife and forestry management has changed its focus to the management of biological diversity, and concern has increased over the adverse effects of fragmenting once-continuous forest landscapes (e.g., avian nest predation [Yahner et al. 1989]). Fragmentation has also been hypothesized to have effects on long-term forest productivity in temperate coniferous forests. For example, it may be having a detrimental impact on populations of mycophagous (fungus-eating) rodents. These rodents are key dispersal agents for the mycorrhizal fungi that have a close symbiotic relationship with many coniferous tree species (Maser and Trappe 1984).

Most studies of the effects of fragmentation on animal species have been conducted in landscapes already dominated by agricultural activities (Opdam 1991). In these cases the nomenclature used to describe the fragmented environment is fairly simple: isolated patches of forest are called fragments; they are surrounded by the agricultural matrix and are sometimes connected by corridors.

In landscapes dominated by forests, fragmentation occurs in three phases (Harris 1984; Franklin and Forman 1987; Spies et al. 1994). Initially the landscape is perforated by the cutting units. In this stage the uncut forest remains connected; habitat loss occurs through the removal of forest and the creation of edge conditions, but the forest remains essentially unfragmented because the uncut forest is still contiguous. As the perforation increases, a point is reached when the cutover areas begin to coalesce and the remaining forest is fragmented into isolated patches. Continued harvesting finally results in the defragmentation of the area by forming a new matrix of harvested forest. In reality, this process is complicated by the regrowth of the harvested forest. Because harvesting and regrowth occur simultaneously, the extent of the fragmentation will depend on the age at which the regrowing forest assumes the characteristics of the original forest (assuming that the management regime allows it to do so) and the rate at which the original forest is harvested.

Saunders et al. (1991) have reviewed in some detail the literature related to the characteristics of fragmented ecosystems. Their information is summarized here, primarily from the point of view of recently fragmented forested landscapes, without reference to original publications. The reader is directed to Saunders et al. (1991) for more detail and for the original references.

Fragmentation produces remnant vegetation patches surrounded by a matrix of different vegetation. The two primary effects are changes in the microclimate of the remnants and isolation of the remnants. Thus, fragmentation results in physical and biogeographic changes.

1   Changes in microclimate
   (a) *Radiation fluxes.* In general, if the forest was dense, the cleared areas will have higher daytime and lower nighttime temperatures. This effect will extend into the remnant patches, and the distance it extends will be strongly influenced by the orientation of the edge with respect to the sun. The impact of this is unclear. It is speculated that the edge may be invaded by shade-intolerant understorey species. This may have a beneficial effect if these plants 'seal off' the edge. Changes in plant communities at the edge and increased radiation load (desiccation) may alter foraging opportunities for herbivores.
   (b) *Wind.* When air flowing over one vegetation type encounters another vegetation type, the upper part of the wind profile initially retains the characteristics of the previous vegetation type while the lower part of the profile takes on the characteristics of the new vegetation type. It is estimated that a minimum of 100-200 times the height of the vegetation is required for the wind profile to fully equilibrate. The implications of this for plant gas exchange have not been studied but could be significant. A more obvious effect is

      direct physical damage to trees at the edge of a recently created remnant, either through wind pruning or windthrow.

(c) *Water flux.* Although Saunders et al. (1991) discuss water flux issues at length, their discussion is more related to the conversion of forested land to agricultural land (or young second growth) rather than to the impact of fragmentation on water flux. In British Columbia, of course, forest harvesting can have dramatic influences on hydrological regimes if it is not done properly. This is a complex process involving such factors as climate and topography (slope position of the remnant stands) as well as the degree of fragmentation of the forest.

2   Isolation. Fragmentation is hypothesized to have two important consequences for biota: the total area of habitat is reduced, and the remnant patches are isolated to varying degrees. As discussed by Saunders et al. (1991), the factors that are believed to have an impact on the biotic response are:

(a) *Remnant size and shape.* The most obvious determinant of the biotic response of remnant patches is their size. At one extreme, a large remnant would be an unfragmented 'mainland.' At the other extreme, a small remnant would contain insufficient habitat to meet the needs of a single individual of the species of concern. Between these extremes, the size of the remnant will determine the amount and, to some extent, the nature of the available habitat. This will influence the species composition that can persist in the remnant. In smaller remnants, the shape will significantly influence the amount of habitat that is not affected by edge characteristics.

(b) *Time since isolation.* It is believed that upon isolation, remnant patches will temporarily contain more species than they can support, for a variety of reasons (see below). The loss of species (termed the *relaxation effect*) will depend on the time since isolation. Short-lived species will probably become locally extirpated more rapidly than long-lived species.

(c) *Distance from other remnants or an unfragmented block.* Other remnants or an unfragmented block can act as sources of immigration for a remnant patch. The degree to which this will be an effective process will depend on the isolation of the remnant and on the dispersal abilities of the species in question.

(d) *Connectivity.* Connectivity is discussed at length in Chapter 3. Suffice it to say here that connectivity can be viewed as the opposite of fragmentation, and should therefore reduce the impact of fragmentation. There appear to be two principal issues: (1) Are there some threshold values for connectivity (such as width, length, or composition) below which the connections among remnants have no value? and (2) What is the 'natural' level of connectivity in the ecosystem that management should strive for?

(e) *Nature of the surrounding habitat.* Dissimilarity between the remnant and the surrounding habitat will affect the ability of dispersing organisms to move from remnant to remnant across the surrounding habitat. The remnant patch will also become populated by species that are not native to the remnant. In forested landscapes, this changes as the altered forest ages (Saunders et al. 1991).

The concern over forest fragmentation was brought to the forefront by Harris (1984) and has its theoretical basis in the island biogeography theory of MacArthur and Wilson (1967). Simply put, smaller oceanic islands have fewer species than larger oceanic islands, and this size effect is modified by the distance of an island from a mainland. If old-growth forest remnants in an 'ocean' of second growth are analogous to oceanic islands, we can expect smaller, more isolated remnants to support fewer species than are found in the unfragmented 'mainland.' For any given species, this relationship is known as an *incidence function* (Wilcove et al. 1986) – that is, the probability of a species occurring in a remnant has some defined relationship with the size of the remnant, and there should be a size limit below which the species will not occur. For the suite of species in a remnant, the relationship is known as a species-area curve (Coleman 1981): $S = CA^Z$, where S is the number of species, A is the area of the patch, C is a scaling constant that varies with location and taxa, and Z is the rate at which species numbers increase with area.

On oceanic islands, observed values for Z vary from 0.25 to 0.50 (MacArthur and Wilson 1967). For Z in this range of values, twice the number of species will require between 4 and 10 times the area. Naturally, area alone cannot completely predict the number of species that will be present on an oceanic island or in a fragmented remnant. Other key factors are the availability of essential resources (food, water, shelter) and the presence of competitors and predators (Usher 1985).

Predictions about the persistence or loss of a species in a fragmented landscape require the use of a theory of the dynamics of metapopulations, or 'a population of populations that go extinct locally and recolonize' (Levins 1970). Metapopulation dynamics are the combined dynamics of the subpopulations (as determined by the impacts of natality and mortality) and the among-fragment dispersal flow (as determined by immigration and emigration). The dynamics of the local subpopulations are affected by the quality and amount of habitat (patch size), stochastic events (such as adverse weather), and other local disturbances (such as nest predation) that can be affected by the nature of the surrounding environment. The dynamics among subpopulations are affected by the relative degree of isolation of the patches and the abilities of the species in question to disperse across the intervening matrix (Opdam 1991).

The theory of island biogeography has been reasonably well validated for oceanic islands (e.g., Schoener and Schoener 1983). The results for fragmented habitat islands in continental environments are more equivocal. The best way to study this topic would be to conduct experimental fragmentation, where the unfragmented forest is surveyed and then deliberate fragmentation is designed to test the theory. Two such experiments are under way, those of Lovejoy et al. (1986) and Margules et al. (in press). In both cases the results are equivocal, probably because the experiments have not been going on long enough to clearly demonstrate what effect fragmentation has.

There are two other approaches to the study of fragmentation. The first is to conduct surveys of fragments over a time series to determine whether or not species are being lost (e.g., Fritz 1979). The difficulty with this approach is that it is impossible to determine whether the observed local extinctions or recolonizations might have occurred in a similar area of unfragmented forest. The second, more common approach is to attempt to correlate the occurrence of species with one or more spatial features of the landscape related to fragmentation (usually patch size, degree of isolation, and time since isolation) at a single point in time (e.g., Schieck 1994). Multi-year studies are often involved, but there is no attempt to infer a rate of loss through time. This approach can be criticized because one cannot assume that the year(s) in which the study occurred are representative, or that the observed correlations reflect causal relationships, or that the observed distributions are in equilibrium with the degree of fragmentation (Opdam 1991).

Opdam (1991) reviewed a large body of literature on the effects of forest fragmentation on bird populations in agricultural landscapes. He found that the studies based on time series demonstrated that local extinction rates were related to fragment size and that recolonization rates were related to the degree of isolation of the fragments. The studies that correlated species occurrence with spatial features of the landscape demonstrated that fragment size, position in the landscape, and the presence of corridors were important. He cautions that 'there is a great need for studies in naturally fragmented landscapes as well as for studies focusing on other, less predictable, habitat types' (Opdam 1991).

With one notable exception (Kattan et al. 1994), forested environments undergoing fragmentation appear to fall into Opdam's (1991) 'less predictable' category. Kattan et al. (1994) demonstrated a clear loss of bird species in the highly fragmented forests of the western Andes of Colombia over the last 80 years. They stated that this was probably an effect of fragmentation because the most vulnerable species (understorey insectivores and large canopy frugivores) were those that were thought to need forest interior conditions. However, they cautioned that it was difficult to come to firm conclusions without a baseline of data from unaltered forest.

There is substantial speculation about the impact of fragmentation on the biota of forested environments (Harris 1984; Spies et al. 1994), and some authors have been able to demonstrate indirect effects such as increased nest predation (Small and Hunter 1988). Although there appears to be little evidence of direct effects of fragmentation on populations, there is a sizable effort to model such effects (e.g., Daust 1994; Gustavson and Crow 1994).

Rosenberg and Raphael (1984) studied the impact of fragmentation on vertebrates in northern California Douglas-fir forests. They found 'the composition of birds, small mammals, reptiles, and amphibians to be very similar in all classes of forest patch size and insularity' and that 'the majority of species we studied exhibited no negative responses, either to reduced patch size or to increased length of adjacent clearcut edge.' The species that showed the strongest negative responses to fragmentation were fisher (*Martes pennanti*), gray fox (*Urocyon cinereoargenteus*), Spotted Owl (*Strix occidentalis*), and Pileated Woodpecker (*Dryocopus pileatus*).

Lehmkuhl et al. (1991) investigated the effect of fragmentation on vertebrate species in the southern Washington Cascade Range. They found that bird species richness actually increased in smaller patches. They attributed this to the 'conventional edge effect' and to the 'packing effect,' where old growth-associated species were packed into the remaining fragments. They found no relationship between small mammal richness and abundance and fragmentation. No consistent or strong relationships between amphibian richness and abundance and levels of fragmentation could be found. Rosenberg and Raphael (1984) and Lehmkuhl et al. (1991) believed that their results could be explained by the fact that the impact of fragmentation in these areas was relatively recent (populations have not equilibrated) and fragmentation had not progressed as far as in the extremely fragmented agricultural environments of eastern North America.

Schieck (1994) studied the relationships between the abundance of small mammals and patch size within fragmented forests on Vancouver Island. The only statistically significant relationship that he found was that bat abundance was *negatively* related to patch size, contrary to the predictions of island biogeography theory. None of the other groups of species showed a significant relationship. Schieck (1994) concluded that his results were preliminary at best because the effects of time since isolation, degree of isolation, and ability of the species to survive in the intervening, harvested forest were not considered.

Mills (1994) found species-specific probabilities of capturing small mammals at varying distances from the edges of old-forest fragments in Oregon.

Deer mice (*Peromyscus manniculatus*) were more likely to be captured near the edge, California red-backed voles (*Clethrionomys occidentalis*) were more likely to be captured away from the edge, and Townsend's chipmunk (*Eutamias townsendi*) showed no response to distance from the edge.

Margules et al. (in press) concluded that 'it is not yet possible to generalize about the responses of species to habitat fragmentation. Management of fragmented ecosystems to maintain biological diversity should be directed at populations of species because different species respond differently.' This emphasis on individual species does not preclude an 'ecosystem management' approach to forest management. The authors were simply saying that when we design and monitor forest management plans, the fate of individual species is more important than the number of species that will be present.

Spatial Patterns in Forested Landscapes

Description of Landscape Patterns

An understanding of landscape patterns must begin with a description of those patterns. The structural elements of a landscape (usually represented as polygons on a land cover map) can be described in terms of type (origin, disturbance regime, physiognomy, species composition), size, shape, heterogeneity, and boundary characteristics. Landscapes can be characterized in terms of dimensions (extent) and the relative abundance and spatial arrangement of the structural elements (Forman and Godron 1986), as well as the association of these elements with other aspects of the environment such as topography and hydrology.

Methods of describing landscape patterns fall into two broad (and not mutually exclusive) categories:

- 'Landscape indices' are frequently calculated to compare and contrast landscapes in different places and in the same place at different times (change detection).
- Hypothetical landscapes are created using conceptual models; they are then compared to real landscapes to detect (and, it is hoped, explain) significant differences.

The simplest way of describing landscape structure is to calculate one or more of the vast array of available 'landscape indices.' These indices can be categorized based on the feature of the landscape structure that they measure (modified from Rogers 1993):

1   Dimensions of individual elements
    (a) area, perimeter, average size

(b) shape as measured by a variety of indices usually based on the relationship between perimeter and area, such as fractal dimension (Sugihara and May 1990) or shoreline development (Wetzel 1975:31).

2   Landscape composition

(a) abundance of patches measured as density, percentage of the area in patches, and so on

(b) diversity (information theory) indices such as richness, evenness (Romme 1982), dominance, and diversity (O'Neill et al. 1988)

(c) edge density reported as the length of various types of edge per unit area (Spies et al. 1994).

3   Spatial arrangement of elements

(a) measures of the distance among patches, such as nearest-neighbour distances (Clark and Evans 1954; Ripple et al. 1991) and a variety of dispersion and contagion indices (e.g., Forman and Godron 1986; O'Neill et al. 1988; Ripple et al. 1991)

(b) measures of landscape connectivity, such as linkage intensity (Forman and Godron 1986) and its converse; fragmentation, such as GISfrag (Ripple et al. 1991); and size of matrix patches (Ripple 1994; Spies et al. 1994).

Literally scores of landscape indices fall into these three categories (reviews by O'Neill et al. 1988; Turner 1989; Rogers 1993; and Riitters et al. 1995). Many indices can be calculated using stand-alone computer software (McGarigal and Marks 1995), add-ons to Geographic Information System (GIS) packages such as GRASS (Baker 1994), or existing features of more sophisticated commercial GIS packages. Most authors use a small subset of the indices to describe the landscapes they study; in most cases they do not provide any rationale for choosing one subset over another.

O'Neill et al. (1988) found that three indices (fractal dimension, dominance, and contagion) were useful for discriminating among major landscape types such as urban coastal, mountain forest, and agricultural areas in the eastern United States by monitoring changes in landscape pattern through time. Recently, Riitters et al. (1995) conducted multivariate factor analysis of 55 landscape metrics calculated for 85 U.S. Geological Survey Land Use Data Analysis maps from across the continental United States. They used a two-step process to select a subset of 6 landscape metrics that they felt best represented the differences among the maps. First, the 55 metrics were grouped based on high correlation (> 0.9), and 1 metric was selected from each group based on its apparent normality. This eliminated 29 metrics, primarily because many were based on the same underlying variables (area and perimeter). A factor analysis was then performed, and the metric with the highest loading for each resulting factor was chosen as the factor's 'representative.' The authors concluded that 5 metrics could be used to adequately describe the differences among the maps:

- average patch perimeter-area ratio
- Shannon Contagion (O'Neill et al. 1988; Li and Reynolds 1993), a measure of how often one cover type is beside another cover type
- average patch area, normalized to the area of a square with the same perimeter
- patch perimeter-area scaling; a value ranging from 0 for linear patches to 1 for square patches and that is also sensitive to patch area number of land cover types.

Riitters et al. (1995) cautioned that this single analysis had not identified all the important dimensions of landscape pattern, particularly if maps of different scales were to be examined. The did say that the analysis represented a starting point from which to compare other maps.

While landscape indices allow us to quantify various aspects of patterns in landscapes, they do not allow us to resolve two fundamental issues in the field of landscape ecology. First, 'wildlife ecology and behaviour may be strongly dependent on the nature and pattern of landscape elements, but few precise measurements relating these changes to the landscape spatial alteration have been made' (Ripple et al. 1991). To put it more bluntly, 'factor analysis can show which metrics appear to measure image texture and contagion, it cannot indicate whether contagion is worth measuring at all' (Riitters et al. 1995). It is clear, even to the most casual observer, that computer technology has allowed our ability to 'measure' landscapes to far outstrip our understanding of what those measurements mean (Berry 1995).

Second, landscape indices allow us to quantify differences among different areas or the same area at different times. They do not allow us to say which of those differences is significant, from either a statistical or an ecological point of view. Some progress in this area has been made primarily through the use of 'neutral landscape models.' Gardner et al. (1987) first proposed neutral landscape models as the 'null hypothesis' of landscape pattern against which real landscapes could be compared. The models generated random landscapes with two types of 'habitat' (susceptible to disturbance and not susceptible to disturbance), using percolation theory as the basis. A probability of being in each type of 'habitat' was provided and cells were then randomly assigned based on the probability. The models created patterns that were neutral to the physical and biological processes that create real landscapes. A further refinement was added by Gardner and O'Neill (1991), who introduced contagion as a factor that could be used to create neutral landscapes with contiguous patches that retain the same proportion of susceptible habitat.

Gardner et al. (1987) used neutral models to help define appropriate scales for landscape studies. Turner et al. (1989) examined the characteristics of landscapes created by disturbances with different probabilities of initiation and spread. Finally, Turner et al. (1994) compared the results of simulated disturbances in Yellowstone National Park with those from neutral landscapes with the same proportion of susceptible habitat. They stated that significant departures of the observed results on real landscapes from the expected results on neutral landscapes could be used to generate and test hypotheses about landscape function.

Examples of Forested Landscape Patterns
There are relatively few published studies of landscape patterns in forested environments in North America. All of these studies, with one exception (Mladenoff et al. 1993), are from the Pacific Northwest. They attempt either to describe the differences between 'unaltered' landscapes and landscapes that have been affected by forest harvesting, or to characterize the change in landscape pattern over recent times. One study (Ripple 1994) describes the 'pre-European' landscape in western Oregon. Several of the more useful studies are summarized below in chronological order of publication.

Harris et al. (1982) examined forest landscape patterns in the Cascade Range of western Oregon and attempted to draw some conclusions about the impact of forest harvesting on wildlife. They found that there had been a shift from harvesting low-elevation forests in the 1940s to harvesting higher-elevation forests since the 1970s. The percentage of area remaining in unharvested forest varied widely depending on the land owner and on the National Forest for lands managed by the U.S. Department of Agriculture (USDA). The authors also noted that there appeared to be a reduction in the size of remaining old-growth stands as well as isolation of such stands. In the Suislaw National Forest, remaining old-growth stands had a median size of 13.6 ha and an interpatch distance approaching 18 km. Based on a literature review of species habitat requirements, they recommended that retained patches of old growth be distributed throughout the elevational range, and that their size should be proportional to their degree of isolation. The authors also expressed substantial concern about the amount of 'middle-aged' forest that would be created by the historical harvesting pattern.

Ripple et al. (1991) examined the changes that occurred in the forest landscape patterns on National Forest land in the Cascade Range of western Oregon between 1972 and 1987. Fifteen rectangular 'landscapes' of 1,750 ha each were randomly chosen from the study area. The authors used aerial photographs and existing land cover mapping to classify the landscapes into harvested patches and the unharvested matrix. They calculated a variety of landscape indices for each landscape in both years, and compared the results using non-parametric statistics. They found that patch size and perimeter decreased and patch density increased. Interior forest area decreased by twice the amount that the area of patches increased. Their measure of fragmentation (GISfrag) indicated that fragmentation had increased. The amount of edge also increased dramatically. These findings were consistent with increased harvest levels and the application of USDA Forest Service policy that promoted small, dispersed cutblocks. The authors concluded that the increase in edge was probably beneficial for deer and elk but detrimental for interior bird species such as Spotted Owl, Townsend's Warbler, and Pileated Woodpecker. They noted that this effect may be relatively hard to detect because fragmentation was a relatively recent phenomenon.

Mladenoff et al. (1993) compared landscape structure in an 'unaltered' (old-growth) landscape and an 'altered' (heavily harvested) landscape in Wisconsin (9,600 ha and 7,200 ha in size, respectively). Their goal was to determine whether structural features at the landscape level could be used to distinguish altered from unaltered landscapes. They believed that if so, the information could be used to direct restoration efforts for altered landscapes. They found that the harvested landscape had significantly more small patches and fewer large, matrix patches than the old-growth landscape. The forest patches in the harvested landscape were simpler in shape (lower fractal dimension) than in the unharvested landscape. The unaltered landscape showed significant changes in patch shape across the size range of patches present. The differences were hypothesized to indicate characteristic difference in patch initiation processes operating at different spatial scales. These differences were lost in the harvested landscape. The harvested landscape was more fragmented than the unharvested landscape: it was characterized by higher isolation of patches and higher edge-to-interior ratios. Some relationships relating to the juxtaposition of landscape elements that appeared to be characteristic of the unaltered landscape (such as old-growth hemlock next to lowland conifers) were absent in the altered landscape. These relationships may be important for both the landscape elements (the old-growth hemlock) and the plants and animals that are adapted to using them.

In general, there is a suggestion that human activities simplify landscapes (Turner and Ruscher 1988). Mladenoff et al. (1993) found that this depended on the scale of observation and the parameter that was measured. Patch complexity was lower on the disturbed landscape, but there were more patches and ecosystem types. The authors cautioned that this apparent increase in landscape diversity was 'an artifact of human disturbance that should not be viewed as a management goal.' The effect was similar to the increase in species diversity that can occur at local levels following disturbance, an increase that results from the creation of habitat for edge species and species associated with early seral stages (Ricklefs 1987).

Examples of a landscape altered by forestry (top) and an unaltered landscape (bottom). (Photographs by Joan Voller)

Ripple (1994) described the spatial patterns of old forests that occurred in western Oregon prior to large-scale forest harvesting. Using 1933 forest cover maps for five forest survey units that had not yet experienced large-scale logging, map polygons were assigned to coniferous forests with large and small trees and a variety of 'non-forested' types. The area, but obviously not the spatial distribution, of old growth was estimated using an estimated minimum age to reach the large conifer class (depending on local site conditions) and an expected negative exponential age distribution model. Ripple (1994) estimated that between 64% and 80% of the area of the units was covered by the large conifer class, with the lowest percentage in areas that had been most affected by large fires during the 1840s. Overall, 71% of the area was in the large conifer class, and 89% of this was connected in the form of one large patch that extended throughout most of Oregon's forest land except the north coast. Ripple (1994) estimated that between 33% and 46% of the units were covered by forests older than 200 years. They also found a direct positive relationship between the distance from rivers and the proportion of the area in the large conifer class, and attributed this to anthropogenic influence, noting that Native Americans, European settlers, and early sawmill operations were all commonly associated with river valleys for a variety of reasons.

Spies et al. (1994) examined the changes that have occurred over a large (2,589 km²) area in Oregon. They categorized satellite imagery for four dates between 1972 and 1988 into closed-canopy forest (largely unharvested forest) and 'other types' (largely clearcut openings). A variety of landscape indices were calculated and compared over time and by low and high elevation and private and public land administration. Closed-canopy forests decreased from 71% of the study area to 58%. This decrease was greatest on private forest land and lowest on public wilderness and reserve lands. Overall, the disturbance rate was 1.19% annually, corresponding to a rotation age of 84 years. Edge density increased from 1.9 to 2.5 km per square kilometre. This increase was most dramatic on public forest land from 1984 to 1988. Correspondingly, interior forest area (as determined by calculating the area of forest that was more than 100 m from a clearing) decreased. By 1988, only 12% of the private land and 43% of the public land was in interior forest.

Interior habitat patches became smaller and more numerous over the study period. In 1972 there was a single large patch (103,608 ha) of interior habitat, and over 50% of the study area was in patches greater than 1,000 ha. By 1988 the largest single patch was 21,018 ha and the mean interior patch size had declined from 160 ha to 62 ha. Spies et al. (1994) concluded that the current disturbance rate was several times more frequent than the natural disturbance regime for the area. They also stated that their study results suggested that cutting rate could have a greater effect on the amount of edge and interior forest in a landscape than cutting pattern. A comparison of subareas on private and public land with similar cutting rates showed that the private land had more interior forest because of the larger cutting units that are employed. However, when the landscape as a whole was examined, private lands had less interior habitat and more edge because of the more rapid cutting rates. The authors concluded that while idealized landscape and reserve design systems would be useful as long-term goals, 'for the interim, landscape managers will need to practise the art of working with the pattern of what they have inherited from nature and previous managers.'

Creating Landscape Patterns with Forestry
Using a simple conceptual model, Franklin and Forman (1987) examined the effects of creating landscape patterns with forest harvesting. They noted that many specific variables were important from an economic and ecological point of view (such as patch size or road location), but that 'the pattern created on the landscape by a sequence of cutting operations may be more critical than these specific variables.' They focused on the effects of the 'checkerboard' cutting pattern of small, dispersed cutblocks that was in vogue on National Forest lands at the time. They found that:

- The length of edge rose linearly until 50% of the area had been cut, and then began to drop.
- The amount of forest interior was related to the amount of edge and the opening size. For example, if the edge width was 160 m (i.e., a 160 m band of forest had an edge influence) and the opening size was 10 ha, there would be no forest interior when 50% of the area had been cut.
- The size of the matrix patches did not decrease until 30% of the area had been cut, but those patches did become more porous.
- When between 30% and 50% of the area had been cut, the size of the matrix patches decreased rapidly.
- The interpatch distance was zero until 50% of the area had been cut, because all the patch corners touched another corner. Interpatch distance then began to increase.

The patterns for the cutover patches mirrored the patterns for the unharvested forest, and the patterns after 50% of the area had been cut mirrored the patterns up to 50% cut. Franklin and Forman (1987) also showed how these patterns would be altered by using a number of simplistic 'alternative' cutting regimes. They noted that real landscapes in the Pacific Northwest deviated from the model outcome: there was a tendency to cut adjacent patches early in the cycle because of blowdown and insect damage salvaging as well as to reduce road costs. They concluded that the patterns created by small, dispersed cutblocks was not particularly desirable. In fact, it has had detrimental effects on windthrow hazard, susceptibility to forest pests, forest fire ignition hazard, mass wasting, and the persistence of species that require forest interior conditions. Other patterns resulting from the progression from one or several nuclei do not have these negative effects, but they require the reservation of large patches for forest interior species and for 'amenity' values.

Li et al. (1993) extended Franklin and Forman's (1987) work with the use of a simulation model to more explicitly describe the effects of different cutting regimes on the resulting forest landscape pattern. They simulated the effects of five cutting regimes: (1) random patches, (2) maximum dispersion of patches, (3) staggered settings (maximum dispersion with larger units), (4) partial aggregation (cutting in one portion of the unit until adjacency constraints are reached), and (5) progressive clearcutting. They calculated four landscape indices: (1) edge density, (2) area weighted shape index, (3) 'patchiness' ([boundary length × dissimilarity score]/total boundary), and (4) interior area fragmentation (interior area/total forest area).

They found that fragmentation increased or fluctuated most with the maximum-dispersion, random, and staggered-setting regimes, less with partial aggregation, and almost not at all with progressive clearcutting. Edge density and patchiness peaked around 50% cut for maximum dispersion and increased slowly to 70% for partial aggregation. The random and staggered-setting regimes had a less dramatic impact on edge density and patchiness. Interior habitat initially decreased rapidly for random patches, maximum dispersion, and staggered settings, and then levelled off at 40-60% of the area cut. The variation in different cutting patterns was usually not observable until 25-30% of the area had been cut (50% in some cases). Li et al. (1993) therefore concluded that fragmentation studies conducted before these thresholds had been reached may not be appropriate. They noted that using cutting patterns that increased cutting unit size (progressive clearcutting) or the density of cutblocks (partial aggregation) decreased fragmentation in the short term but would have negative impacts on other factors, such as forest recovery and peak stream flow. A balancing of objectives was therefore required.

Li et al. (1993) believed that it may be possible to change the course of fragmentation before it reached the threshold after which its effects would be observed (25-50%). However, Wallin et al. (1994) simulated switching from a dispersed cutblock regime to an aggregated cutting regime and found that the switch produced little change in indices such as edge length and interior habitat, even if done as early as 20 years after dispersed cutting had begun. They found that an initially dispersed pattern remained even after an aggregated cutting regime was adopted. This was because the initial units were not eligible for harvesting for 100 years and therefore had to be cut around. 'Erasing this [an initial dispersed pattern] (even after only 20 yr) and establishing a new one requires a substantial change in the individual rules (constraints) that govern the process.' Wallin et al. (1994) found that no rule changes were completely successful; the only adequate method of erasing an initially dispersed cutting pattern was a long-term moratorium on harvesting.

'Some landscape level patterns are highly resistant to change when driven by deterministic disturbance processes governed by simple rules ... Hence, pattern analysis may provide only limited information about changes in these disturbance rules' (Wallin et al. 1994). Baker (1994) came to similar conclusions regarding the restoration of landscapes altered by fire suppression. He concluded that the often recommended procedure of introducing small prescribed fires would further alter, rather than restore, the 'natural' landscape pattern. His simulations of landscape patterns through time in the Boundary Waters Canoe Areas in eastern North America indicated that restoration could be achieved simply by reinstating the disturbance regime that existed before fire suppression. Although some unusually large fires can be expected initially because of fuel build-up, these fires simply hasten the return to the natural pattern.

There is a spectrum of methodologies for designing the patterns that forest management will create at the landscape scale (5,000 to 50,000 ha). One end of this spectrum is characterized by the use of various modifications to traditional linear programming techniques that attempt to incorporate landscape pattern objectives such as cutblock size and adjacency rules as 'spatial constraints.' Examples of this approach are Monte Carlo Integer Programming (Nelson et al. 1991), Simulated Annealing (Lockwood and Moore 1993), and Mixed Integer Programming (Hof et al. 1994). The vast body of literature on this subject is beyond the scope of this chapter. Suffice it to say that these methods attempt to generate a 'nearly optimal' solution to the problem of harvest scheduling in the face of spatial constraints and predefined objectives. Hof et al. (1994) point out that 'the problem of simultaneously optimizing land management for wildlife habitat over space and time is enormously complex ... This is obviously an exploratory effort, and a great deal more work is called for.' However, Holgrem (1995) is not optimistic about the gains that can be made with these kind of methods. 'Long term planning must be based on a more simple and flexible model that allows for changes. Since many long-term plans are *de facto* revised every 5 to 10 years, planning can be characterized as an iteration over time. In this context, the rotation period becomes strange, as no reliable stand-based planning can be made for time periods long enough to incorporate several stand rotations ... 100 year harvest schedules cannot be more than a theoretical vision' (Holgrem 1995).

In this photograph taken on northern Vancouver Island, recent clearcuts show a developing landscape pattern created through forestry. There is a spectrum of methodologies for designing the patterns that forest managment will create at the landscape scale. (Photograph by Joan Voller)

The other end of the spectrum is well represented by Diaz and Apostol (1992). This approach has its roots in the field of landscape architecture and represents a very human-oriented process. The forest landscape is 'analyzed' by an interdisciplinary team, which then 'designs' what amounts to a detailed land-use plan where the forest pattern for each unit in the plan is specified. This approach does not result in a harvest schedule; its result is more like a detailed operability map. This type of integrated or 'holistic' planning is advocated by others as well (Cissel et al. 1994; Scientific Panel 1995). As pointed out by Cissel et al. (1994), elements of both the 'engineering' and the 'architectural' approach to landscape design will be needed to plan adequately.

Both approaches to landscape-level planning require that objectives for the landscape be set. The 'architectural' approach usually includes an objective-setting step. The 'engineering' approach usually requires that some principal objective (e.g., timber flow, wildlife habitat) be set initially, and other objectives are treated as constraints in the models. Deciding what we want means answering the questions how much, when, and where. For specific resources that are important economically, socially, or ecologically, these questions probably should be answered on a resource-by-resource basis. However, because of limited knowledge and analytical capability, it is not possible to simultaneously answer these questions for all resources. Thus an 'ecosystem management' approach is often suggested (Forest Ecosystem Management Assessment Team 1993). The specification of objectives for this approach is usually done in terms of the natural disturbance regimes (Province of British Columbia 1995b) or the range of natural variability in the ecosystem attributes (Swanson et al. 1993).

Natural variability is defined as the composition, structure, and dynamics of ecosystems before the influence of European settlers. It is characterized by the natural range of ecosystem conditions and the variety of seral classes that result from the disturbance regime (frequency, severity, spatial arrangement) that affected the area. It is assumed that deviation from historical conditions puts species at risk because they are adapted to those (Holocene) conditions. The intent is not to return to a 'wilderness' condition or any single, pre-existing condition but rather to bring existing conditions within the natural range, or to use the natural landscape patterns as a point of reference in examining the effects of alternative management scenarios.

This approach is still in its infancy in the United States and several problems need to be worked out:

- It is difficult to specify what the range of natural variability was because the record of disturbance is frequently too short, the impact of Native and European disturbance is confounding, and the frequency and severity of disturbances vary over time and space (refer to Chapter 1 for more information on natural disturbance ecology).
- A great deal has changed since the pre-European era (e.g., introduced exotics, climate), and selecting a range of conditions at any previous time may be both biologically and socially arbitrary (is it what we want?).
- The disturbance regime created by management may interact with engineered structures or climate change to produce results outside the range of natural variability (Swanson et al. 1993).

Finally, Swanson et al. (1993) point out that the 'range of natural variability of ecosystems and landscapes is likely to differ in some important respects from the conditions desired by society for many lands, perhaps even wilderness.' Balance between what is perceived to be ecologically correct and what is socially desirable is therefore needed in setting objectives.

Research Needs

Research is needed to examine pattern and scale in ecology. According to Levin (1992), the two fundamental and interconnected themes in ecology are the development and maintenance of a spatial and temporal pattern and the consequences of that pattern for the dynamics of populations and eco-systems. Levin (1992) outlines a research program for the examination of pattern and scale:

1   Measures are needed to describe pattern so that criteria can be given for relating pattern to cause.
2   Cross-correlation examination can provide initial suggestions as to mechanisms.
3   Theories must be developed that catalogue the possibilities and capture emergent phenomena that arise from the collective behaviour of smaller-scale processes.
4   Experiments must be conducted to determine what is probable.

We have begun to develop measures for describing patterns, but it appears that much of the rest of Levin's (1992) program is being done in an ad hoc fashion, if it is being done at all.

Research is needed to understand the effects of fragmentation on the biota of forested environments. The effects of fragmentation on the biota of forested environments have not been well documented. 'There is a great need for studies in naturally fragmented landscapes as well as for studies focusing on other, less predictable, habitat types' (Opdam 1991). We could benefit from the examination of some extreme cases of fragmentation (e.g., remnant patches in extremely large clearcuts).

Other research issues around the topic of fragmentation include the following:

To what extent does the difference between habitats in a fragmented landscape have an impact on the consequences of fragmentation? That is, the 'islands' of remnant forest are not islands in the oceanic sense. The intervening habitat has some quality for some organisms, and the nature of this habitat will affect the degree of isolation of the remnants.

How to manage for fragmentation at different scales. Like all other issues in ecology, fragmentation is scale-dependent. Can we identify and manage for fragmentation at the scales that are relevant to beetles and grizzly bears?

A great deal of research on forest landscape design and ecosystem management (range of natural variability) is required:

Tools for developing and evaluating forest landscape designs. These include tools for designing forest landscapes, scheduling forestry activities, and assessing costs and benefits for all resources.

Methods to determine and quantify the ranges of natural variability in various ecosystem attributes. The inclusion of a more realistic representation of the forest in the 'neutral models' developed by Gardner et al. (1987) may be helpful in this regard.

Literature Cited

Baker, W.L. 1994. Restoration of landscape structure altered by fire suppression. Conservation Biology 8:763-769.

B.C. Ministry of Environment, Lands and Parks. 1994. Maintaining British Columbia's wildlife heritage. Provincial Wildlife Strategy, B.C. Ministry of Environment, Lands and Parks, Victoria, B.C.

B.C. Ministry of Forests. 1996. Higher level plans: policy and procedures. B.C. Ministry of Forests, Victoria B.C.

Berry, J.K. 1995. Is the GIS cart before the horse? GIS World 8(3):34-38.

Brown, J.H., and P.F. Nicoletto. 1991. Spatial scaling of species composition: body masses of North American land mammals. American Naturalist 138:1479-1512.

Chatfield, C. 1984. The analysis of time series: an introduction. 3rd ed. Chapman and Hall, London.

Chen, J., J.F. Franklin, and T.A. Spies. 1993. Contrasting microclimates among clearcut, edge, and interior of old-growth Douglas-fir forest. Agricultural and Forest Meteorology 63:219-237.

Cissel, J.H., F.J. Swanson, W.A. McKee, and A.L. Burditt. 1994. Using the past to plan the future in the Pacific Northwest. Journal of Forestry 92:30-31.

Clark, P.J., and F.C. Evans. 1954. Distance to nearest neighbor as a measure of spatial relationships in populations. Ecology 35:445-453.

Coleman, B.D. 1981. On random placement and species-area relations. Mathematical BioScience 54:191-215.

Daust, D.K. 1994. Biodiversity and land management: from concept to practice. M.Sc. thesis, University of British Columbia, Faculty of Forestry, Vancouver, B.C.

Davis, J.C. 1986. Statistics and data analysis for geology. 2nd ed. John Wiley and Sons, New York.

Delcourt, H.R., and P.A. Delcourt. 1988. Quaternary landscape ecology: relevant scales in space and time. Landscape Ecology 2:45-61.

Diaz, N., and D. Apostol. 1992. Forest landscape analysis and design: a process for developing and implementing land management objectives for landscape patterns. USDA Forest Service Pacific Northwest Region Publication R6 ECO-TP-043-92.

Forest Ecosystem Management Assessment Team (FEMAT). 1993. Forest ecosystem management: an ecological, economic and social assessment. USDA Forest Service (and other agencies), Portland, Ore.

Forman R.T.T., and M. Godron. 1986. Landscape ecology. John Wiley and Sons, New York.

Franklin, J., and R.T.T. Forman. 1987. Creating landscape patterns by forest cutting: ecological consequences and principles. Landscape Ecology 1:5-18.

Fritz, R.S. 1979. Consequences of insular population structure: distribution and extinction of spruce grouse populations. Oecologia 42:57-65.

Gardner, R.H., B.T. Milne, M.G. Turner, and R.V. O'Neill. 1987. Neutral models for the analysis of broad-scale landscape patterns. Landscape Ecology 1:19-28.

Gardner, R.H., and R.V. O'Neill. 1991. Pattern, process and predictability: the use of neutral models for landscape analysis. Pages 289-307 in M.G. Turner and R.H. Gardner (eds.). Quantitative methods in landscape ecology. Ecological Studies 82, Springer-Verlag, New York.

Greig-Smith, P. 1964. Quantitative plant ecology. 2nd ed. Butterworths, London.

Gustavson, E.J., and T.R. Crow. 1994. Modeling the effects of forest harvesting on landscape structure and the spatial distribution of cowbird parasitism. Landscape Ecology 9:237-248.

Harris, L.D. 1984. The fragmented forest: island biogeography theory and the preservation of biotic diversity. University of Chicago Press, Chicago, Ill.

Harris, L.D., C. Maser, and A. McKee. 1982. Patterns of old growth harvest and implications for Cascades wildlife. Pages 374-393 in Transactions of the Forty-seventh North American Wildlife and Natural Resources Conference. Wildlife Management Institute, Washington, D.C.

Harvey, P.H., and M.D. Pagel. 1991. The comparative method in evolutionary biology. Oxford University Press, Oxford, England.

Hill, M.O. 1973. The intensity of spatial pattern in plant communities. Journal of Ecology 61:225-235.

Hof, J., M. Bevers, L. Joyce, and B. Kent. 1994. An integer programming approach for spatially and temporally optimizing wildlife. Forest Science 40:177-191.

Holgrem, P. 1995. Geographic information for forestry planning. Reports in Forest Ecology and Forest Soils no. 68. Swedish University of Agricultural Sciences, Uppsala, Sweden.

Kattan, G.H., H. Alvarez-Lopez, and M. Giraldo. 1994. Forest fragmentation and bird extinctions: San Antonio eight years later. Biological Conservation 8:138-146.

Lee, K.N. 1994. Greed, scale mismatch, and learning. Ecological Applications 3(4):560-564.

Lehmkuhl, J.F., L.F. Ruggiero, and P.A. Hall. 1991. Landscape-scale patterns of forest fragmentation and wildlife richness and abundance in the southern Washington Cascade Range. Pages 425-442 in L.F. Ruggiero, K.B. Aubry, A.B. Carey, and M.H. Huff (eds.). Wildlife and vegetation of unmanaged Douglas-fir forests. USDA Forest Service General Technical Report PNW-GTR-285. Portland, Ore.

Leopold, A. 1933. Game management. Charles Scribner's Sons, New York.

Levin, S.A. 1992. The problem of pattern and scale in ecology. Ecology 73:1943-1967.

Levins, R. 1970. Extinctions. Pages 77-107 in Some mathematical questions in biology: lectures on mathematics in the life sciences. American Mathematical Society, Providence, R.I.

Li, H., and J.F. Reynolds. 1993. A new contagion index to quantify spatial pattern. Landscape Ecology 8:155-162.

Li, H., J.F. Franklin, F.J. Swanson, and T.A. Spies. 1993. Developing alternative forest cutting patterns: a simulation approach. Landscape Ecology 8:63-75.

Lockwood, C., and T. Moore. 1993. Harvest scheduling with spatial constraints: a simulated annealing approach. Canadian Journal of Forest Research 23:468-478.

Lovejoy, T.E., R.O. Bierregaard, A.B. Malcolm, J.R. Rylands, C.E. Quintela, L.H. Harper, K.S. Brown, A.H. Powell, G.V.H. Powell, H.O.R. Shubert, and M.B. Hays. 1986. Edge and other effects of isolation on Amazon forest fragments. Pages 257-285 in M.E. Soulé (ed.). Conservation biology: the science of scarcity and diversity. Sinauer Associates, Sunderland, Mass.

Ludwig, D., R. Hilborn, and C. Walters. 1993. Uncertainty, resource exploitation, and conservation: lessons from history. Science 260:17-36.

Ludwig, J.A. 1979. A test of different quadrat variance methods for the analysis of spatial pattern. Pages 298-304 in R.M. Cormack and J.K. Ord (eds.). Spatial and temporal analysis in ecology. Statistical Ecology Series vol. 8. International Cooperative Publishing House, Fairland, Md.

MacArthur, R.H., and E.O. Wilson. 1967. The theory of island biogeography. Princeton University Press, Princeton, N.J.

Mandelbrot, B.B. 1983. The fractal geometry of nature. W.H. Freeman, New York.

Margules, C.R., G.A. Milkkovits, and G.T. Smith. (in press). Contrasting the effects of habitat fragmentation on the scorpion *Cercophonius squama* and an amphipod.

Maser, C., and J.M. Trappe. 1984. The seen and unseen world of the fallen tree. USDA Forest Service General Technical Report PNW-GTR-133.

McGarigal, K., and B. Marks. 1995. Fragstats: a spatial pattern analysis program for quantifying landscape structure. USDA Forest Service General Technical Report PNW-GTR-351. Portland, Ore.

Mills, L.S. 1994. Effects of forest fragmentation on small mammals in southwest Oregon. COPE Report 7:6-11.

Mladenoff, D.J., M.A. White, and J. Pastor. 1993. Comparing spatial pattern in unaltered old-growth and disturbed forest landscapes. Ecological Applications 3:294-306.

Nelson, J., J.D. Brodie, and J. Sessions. 1991. Integrating short-term, area-based logging plans with long-term harvest schedules. Forest Science 37:101-122.

O'Neill, R.V., J.R. Krummel, R.H. Gardner, G. Sugihara, B. Jackson, D.L. DeAngelis, B.T. Milne, M.G. Turner, B. Zygemunt, S.W. Christensen, V.H. Dale, and R.L. Graham. 1988. Indices of landscape pattern. Landscape Ecology 1:153-162.

O'Neill, R.V., S.J. Turner, V.I. Cullinan, D.P. Coffin, T. Cook, W. Conley, J. Brunt, J.M. Thomas, M.R. Conley, and J. Gosz. 1991. Multiple landscape scales: an intersite comparison. Landscape Ecology 5:137-144.

Opdam, P. 1991. Metapopulation theory and habitat fragmentation: a review of holarctic breeding bird studies. Landscape Ecology 5:93-106.

Peters, R.H. 1991. A critique for ecology. Cambridge University Press, Cambridge, England.

Province of British Columbia. 1995a. The Forest Practices Code of British Columbia Act and Regulations. Order in Council 429/95. Queen's Printer, Victoria, B.C.

Province of British Columbia. 1995b. Biodiversity guide book. Forest Practices Code guide book. Victoria, B.C.

Ricklefs, R.E. 1987. Community diversity: relative roles of local and regional processes. Science 235:167-171.

Riitters, K.H., R.V. O'Neill, C.T. Hunsacker, J.D. Wickham, D.H. Yankee, S.P. Timmins, K.B. Jones, and B.L. Jackson. 1995. A factor analysis of landscape pattern and structure metrics. Landscape Ecology 10:23-39.

Ripple, W.J. 1994. Historic spatial patterns of old growth forests in western Oregon. Journal of Forestry 92:45-49.

Ripple, W.J., G.A. Bradshaw, and T.A Spies. 1991. Measuring forest landscape patterns in the Cascade Range of Oregon, USA. Biological Conservation 57:73-88.

Robson, P.A. 1995. The working forest of British Columbia. Harbour Publishing, Madeira Park, B.C.

Rogers, C.A. 1993. Describing landscape: indices of structure. M.Sc. thesis, Simon Fraser University, Burnaby, B.C.

Romme, W.H. 1982. Fire and landscape diversity in subalpine forest of Yellowstone National Park. Ecological Monograph 52:199-221.

Rosenberg, K.V., and M.G. Raphael. 1984. Effects of forest fragmentation on vertebrates in Douglas-fir forest. Pages 263-272 in J. Verner, M.L. Morrison, and C.J. Ralph (eds.). Wildlife 2000: modeling habitat relationships of terrestrial vertebrates. University of Wisconsin Press, Madison, Wis.

Saunders, D.A., R.J. Hobbs, and C.R. Margules. 1991. Biological consequences of ecosystem fragmentation: a review. Conservation Biology 5:18-32.

Schieck, J. 1994. Relationships between the abundance of small mammals and patch size within fragmented forests on Vancouver Island. FRDA Research Memo no. 222, Victoria, B.C.

Schimmel, D.S., W.J. Parton, T.G.F. Kittel, D.S. Ojima, and C.V. Cole. 1990. Grassland biochemistry: links to atmospheric processes. Climate Change 17:13-25.

Schneider, S.H. 1989. The greenhouse effect: science and policy. Science 243:771-781.

Schoener, T.W., and A. Schoener. 1983. The time to extinction of a colonizing propagule of lizards increases with island area. Nature 302:332-334.

Scientific Panel for Sustainable Forest Practices in Clayoquot Sound. 1995. Sustainable ecosystem management in Clayoquot Sound: planning and practices. Report 5. Victoria B.C.

Small, M.F., and Hunter, M.L. 1988. Forest fragmentation and avian nest predation in forested landscapes. Oecologia 76:62-64.

Sokal, R.R., G.M. Jaques, and M.C. Wooten. 1989. Spatial autocorrelation analysis of migration and selection. Genetics 121:845-856.

Spies, T.A., W.J. Ripple, and G.A. Bradshaw. 1994. Dynamics and pattern of a managed coniferous forest landscape in Oregon. Ecological Applications 4(3):555-568.

Sugihara, G., and R. May. 1990. Applications of fractals in ecology. TREE 5:79-86.

Swanson, F.J., J.A. Jones, D.O. Wallin, and J.H. Cissel. 1993. Natural Variability – implications for ecosystem management. Pages 89-103 in M.E. Jensen and P.S. Bourgeron (eds.). Eastside forest ecosystem health assessment. Volume 2, Ecosystem management: principles and applications. USDA Forest Service, Northern Region, Missoula, Mont.

Turner, M.G. 1989. The effect of pattern on process. Annual Review of Ecology and Systematics 20:170-197.

Turner, M.G., and C.L. Ruscher. 1988. Changes in landscape patterns in Georgia. Landscape Ecology 1:241-251.

Turner, M.G., and R.H. Gardner. 1991. Quantitative methods in landscape ecology. Introduction. Pages 3-14 in M.G. Turner and R.H. Gardner (eds.). Quantitative methods in landscape ecology. Springer-Verlag, New York.

Turner, M.G., R.V. O'Neill, W. Conely, M.R. Conley, and H.C. Humphries. 1991. Pattern and scale: statistics for landscape ecology. Pages 19-49 in M.G. Turner and R.H. Gardner (eds.). Quantitative methods in landscape ecology. Springer-Verlag, New York.

Turner, M.G., Costanza, R., and F.H. Sklar. 1989. Methods to evaluate the performance of spatial simulation models. Oikos 55:121-129.

Turner, M.G., W.H. Romme, and R.H. Gardner. 1994. Landscape disturbance models and the long-term dynamics of protected areas. Natural Areas Journal 14:3-11.

Urban, D.L., R.V. O'Neill, and H.H. Shugart Jr. 1987. Landscape ecology. BioScience 37:119-121.

Usher, M.B. 1985. Implications of species-area relationships for wildlife conservation. Journal of Environmental Management 21:181-191.

Wallin, D.O., Swanson F.J., and B. Marks. 1994. Landscape pattern response to changes in generation rules: land-use legacy in forestry. Ecological Applications 4:569-580.

Wetzel, R.G. 1975. Limnology. W.B. Saunders, Philadelphia, Pa.

Wilcove, D.S., C.H. McLellan, and A.P. Dobson. 1986. Habitat fragmentation in the temperate zone. Pages 237-256 in M.E. Soulé (ed.). Conservation biology: the science of scarcity and diversity. Sinauer Associates, Sunderland, Mass.

Wilcox, B.A., and D.D. Murphy. 1985. Conservation strategy: the effects of fragmentation on extinction. American Naturalist 125:879-887.

Yahner, J., T.E. Morrell, and J.S. Rachael. 1989. Effect of edge contrast on depredation of artificial avian nests. Journal of Wildlife Management 53:1135-1138.

Yoakum, J., and W. Dasmann. 1971. Habitat manipulation practices. Pages 173-231 in R. Giles (ed.). Wildlife management techniques. Wildlife Society, Washington, D.C.

Zipperer, W.C. 1993. Deforestation patterns and their effects on forest patches. Landscape Ecology 8:177-184.

Zonneveld, I.S. 1979. Land evaluation and land(scape) science. International Training Centre, Enschede, Netherlands.

3
Connectivity
Scott Harrison and Joan Voller

Definitions

Connectivity is an ecological term that describes connections among habitats, species, communities, and ecological processes. Connectivity enables the flow of energy, nutrients, water, disturbances, and organisms and their genes at many spatial and temporal scales (Noss and Harris 1989; Noss 1991). Connectivity for fauna can be measured as the probability that a species will move between patches in the landscape (Taylor et al. 1993). These movements depend on how close the patches are and how well the patches are connected (Taylor et al. 1993). As connectivity between patches is lost, the patches become more isolated and the landscape becomes more fragmented.

In forested landscapes, connectivity describes the linkages among areas of forest cover. In an area where there is no disturbance, the level of connectivity is 100%. (Photograph by Scott Harrison)

Connectivity often is discussed in terms of the mechanisms used to achieve or maintain natural connectivity between patches, namely, corridors or linkages. However, corridors are not always necessary to maintain connectivity. Individual patches that are suitable for a species to move through or live in may constitute connectivity (Merriam 1991a). Corridors may be defined natural features such as riparian habitats along a water course or human-built features such as fencerows. The use of the word *corridor* often implies 'a linear remnant of naturally occurring vegetation that allows movement of individuals or genes between patches of natural habitat' (Harris and Scheck 1991). The following are other definitions of mechanisms for maintaining connectivity.

Corridor, faunal dispersal corridor, movement corridor or conservation corridor: a 'linear remnant' of vegetation, unlike the surrounding vegetation, that connects at least two historically connected patches (Saunders and Hobbs 1991a; Hobbs 1992). A naturally occurring or restored strip of landscape that connects larger patches of similar habitat; the strip functions as a movement route for individuals or for gene flow of native flora and fauna (Harris and Atkins 1991; Harris and Scheck 1991). Any habitat through which an animal has a high probability of moving (Noss 1991).

Convex corridor: a corridor that has greater height than the surrounding matrix, such as a hedgerow, forest in clearcut or agriculture, and shelterbelts (Gates 1991).

Concave corridor: a corridor that is lower in height than the surrounding matrix, such as a powerline, railroad, trail, and highway right-of-way (Gates 1991).

Strip corridor: a corridor that is wide enough to contain interior habitat (Forman 1983; Hobbs 1992).

Line corridor: a corridor that is thin enough to have an edge effect throughout (Forman 1983; Hobbs 1992).

Sink corridor: a corridor that entices animals away from one patch but fails to take them to another patch (Forman 1991; Simberloff et al. 1992).

Riparian or stream corridor: a linear landscape of natural habitat that includes the stream or waterway and the portion of landscape up to and including the level where vegetation may be influenced by high water or flooding and by the ability of soils to hold water (Naiman et al. 1993).

Utility or transportation corridor: a linear strip of development through an area of natural vegetation, such as a trail, railroad, powerline, highway right-of-way, gas pipeline, and so on.

Underpass or tunnel: human-built structures that enable faunal movement across or under utility corridors, such as highways (Simberloff et al. 1992).

Windbreak or shelterbelt: a convex corridor or linear belt of trees in an agricultural setting used to decrease wind and soil erosion (Cable 1991).

Greenway: a linkage of natural or planted vegetation that connects open spaces such as parks, recreational areas, nature reserves, cultural areas, and historic features, and that follows natural (rivers, streams, ridges) or human-built features (roadsides, railroads, canals); trails are not considered greenways but may be part of them (Hay 1991).

Forest Ecosystem Network: linkages are created to maintain the dynamics of natural forest ecosystem processes, provide relatively undisturbed interior habitat, and aid in recolonization, dispersal, and movement (McDougall 1991).

Landscape linkage: corridor within which community and ecosystem processes occur, allowing movements to occur over generations (Csuti 1991).

Biogeographic landbridge: large linkages that aid in the intercontinental movement of complete communities, such as the Isthmus of Panama (Simberloff et al. 1992).

Stepping stone effect, discrete refuges: a series of distinct patches that may act as a corridor for such species as migratory waterfowl (Date et al. 1991; Simberloff 1992).

Background

The concept of corridors was first proposed by Simpson (1936), who used the term to describe situations in which two biogeographically separate regions came close enough to exchange fauna (e.g., the Isthmus of Panama) (Simpson 1940). His view of corridors was on a much grander, continental scale than we generally consider today. Although many ecologists are again pointing out the importance of this continental scale for maintaining connectivity for wide-ranging fauna (Noss 1992). More recent views of corridors as mechanisms to maintain connectivity within a landscape began with Preston (1962). He concluded that a park could not retain the complete inventory of fauna that would occur in a larger area, but corridors between reserves may prevent isolation and total faunal collapse.

The early corridor work was followed by MacArthur and Wilson's (1967) *Theory of Island Biogeography.* This theory states that the number of species on an oceanic island represents a balance between immigration and extinction, and is a function of the island's area and the degree of isolation. MacArthur and Wilson (1967), and later Wilson and Willis (1975), proposed that 'habitat islands,' created through habitat fragmentation, were similar to true oceanic islands. They suggested that the number of species that would reach equilibrium on these habitat islands was a function of the area and the degree of isolation from a 'mainland' source. As an area became more isolated, its natural diversity would decrease until it reached a new, lower equilibrium. Field data collected by Diamond (1975) in New Guinea and Terborgh (1974) on Barro Colorado Island in the Panama Canal supported this hypothesis.

Wilson and Willis (1975) went on to propose the use of corridors as a method to decrease the isolation of habitat fragments and thus decrease the likelihood of losing natural diversity. Wilson and Willis (1975) and Simberloff and Abele (1976) proposed that to preserve populations of unique biota and habitats, it may be better to have several small, connected reserves rather than one large reserve. They argued that species seldom become extinct in all parts of their range at the same time, and there were dangers in putting all individuals of a given species into one refuge. For example, the heath hen (*Tympanachus cupido cupido*) became extinct when a disease common to poultry infected the last remaining birds in their single refuge (Simberloff and Cox 1987). However, Simberloff and Cox (1987) acknowledge that there is a limit to how small a reserve can be. Other studies also have demonstrated that there is a limit to how small and isolated a reserve can be and still retain its natural faunal assemblage. Newmark (1987) studied 14 western North American parks. Thirteen of the 14 parks had lost 43% of their original mammalian assemblage. Only the large area linking several parks, the Kootenay-Banff-Jasper-Yoho, had managed to maintain all its original mammalian fauna.

In 1970, Levins introduced the metapopulation model. The fundamental metapopulation model is an interconnected group of populations where there is interchange of individuals. Some populations may experience local extirpations, but recolonization by connected populations prevents the overall extinction of the species (Merriam 1991a). This theory assumes that (1) more than one mainland source exists, and (2) recolonization is not a rare and limiting factor. Metapopulation theory also incorporates the effects of heterogeneity within and among patches used by different populations (Merriam 1991a).

Metapopulations proved to be an attractive model upon which to base habitat management. Support for the metapopulation model increased, in part, because some ecologists became disillusioned with conservation strategies that conserved species numbers rather than species natural diversity (McEuen 1993). Fragmentation was viewed as having the effect of converting continuous, stable populations into increasingly unstable metapopulations. The conservation movement soon adopted the idea that populations require links among patches, and that the solution for the conservation of many species consisted of multiple reserves linked by corridors. Empirical data, however, reveal that there are few populations that fit the basic model. Harrison (1991) suggests that there are different, more common types of metapopulations: source-sink populations, patchy populations, and non-equilibrium metapopulations. To determine which model best describes a population, one requires an understanding of the population's dynamics, including the role of dispersal, recolonization, or gene flow.

Many studies on the use and importance of corridors have involved small mammals and birds (Bennett 1990; Merriam and Lanoue 1990; Saunders and de Rebeira 1991). Merriam and Lanoue (1990) found that white-footed mice (*Peromyscus leucopus*) preferred fencerows to other landscape types. Data from Fahrig and Merriam (1985) suggested that isolated populations of white-footed mice had lower growth rates and were more prone to extinction than connected populations. Bennett (1990) observed that small mammals frequently used corridors between forest patches and that corridors maintained continuity among isolated populations. Saunders and de Rebeira (1991) observed that as native vegetation decreased, so did native fauna. Their study indicated that linkages between patches of habitat could slow or halt faunal relaxation (decreases in the number of species in an area) caused by fragmentation of the habitat.

A study in the eastern United States (MacClintock et al. 1977) inferred that corridors were important for birds living in a fragmented habitat. MacClintock et al. (1977) indicated that fragmentation was a serious problem, and that it had detrimental effects on neomigrants and interior habitat bird species. The study demonstrated a relationship between habitat tract size and avifaunal composition. There was one exception, where a 35-acre site supported a greater number of bird species than many of the larger preserves; unlike the larger preserves, however, this site was connected by a corridor to a 400-acre site that was connected by corridors to a 10,000-acre forest. The species common in the 400-acre site were also common in the 35-acre site, which, with its corridor, also supported breeding pairs of interior habitat species. MacClintock et al. (1977) concluded that forest species could breed in fragments provided the fragments were attached to larger tracts of forest.

Similar movement patterns have been observed in other species, suggesting that connectivity enhances a population's resilience. Fifty moves by 31 voles (*Clethrionomys glareolus*) were noted between a small patch and large patch of forest along a corridor, whereas no moves were noted between a small patch and large patch of forest of similar distance without a corridor (Szacki 1987). Two mule deer herds (*Odocoileus hemionus*) in the Sierra Nevadas used a specific migration corridor that joined their fragmented seasonal ranges (Kucera and McCarthy 1988). The loss of this migration corridor was equated to the loss of deer. A study of cougars (*Felis concolor*) in the Santa Ana Range of California concluded that cougars used corridors and that the loss of the corridor would guarantee the loss of cougars in isolated areas of their range (Beier 1993).

Still, there are few data on the role of connectivity in landscapes. It is a challenge to implement studies of landscape connectivity and its effect on ecological processes. Ecologists try to balance the need to work at suitable spatial scales with research designs that have an appropriate level of

Female spruce grouse will make seasonal movements of up to 30 km one-way. The level of connectivity in the landscape is probably an important factor affecting where spruce grouse are found and how they get to and from these areas. (Photograph by Scott Harrison)

statistical rigour. Nicholls and Margules (1991) discuss MacClintock et al. (1977) and provide options for study designs intended to test the effectiveness of connections in existing landscapes.

Harrison et al. (in press) is one example of a study that is attempting to achieve statistical rigour at a relatively large spatial scale of research. This study is examining the effects of landscape connectivity on the ecological processes of natality, mortality, movement, and dispersal. After consideration of some ecological and logistical factors, Harrison et al. (in press) are studying 100 ha patches of forest in 400,000 ha of British Columbia's subboreal spruce forest (*Picea glauca* x *engelmannii*). They are quantifying the level of connectivity to each of the patches as low or medium based on the percentage of connections remaining after logging. They also are interested in the effect of increases in population density; therefore, the resulting study layout is a factorial experiment with two factors and two levels. In total, they selected eight forest patches, each 100 ha, to allow for replication of all treatment units. They chose spruce grouse (*Dendragapus canadensis*) as their study species because they felt it had characteristics that made it well suited for a study of landscape connectivity.

As studies continue, so do discussions about ways to address the issue of connectivity. Simberloff and Cox (1987) present many salient points about the use of corridors for conservation efforts in Florida. The situation in Florida differs from that in most other areas of North America in that Florida is facing reconstruction of habitats to provide corridors for species movement rather than maintenance of natural connectivity. Nevertheless, the authors raise points that are applicable elsewhere, such as a concern over the increased exposure to humans and predators of animals using corridors. Another concern is economic: is it wiser to spend money on buying up land for corridors or on purchasing land for large reserves? This debate sparked a series of articles in which Noss (1987) and Simberloff and Cox (1987) presented the advantages and disadvantages of corridors (Table 3.1).

Advocates of corridors argue that linkages should be established or maintained only where such connectivity occurred in the recent past (Harris and Scheck 1991; Saunders and Hobbs 1991a). This approach defuses many of the concerns about corridors opening up dispersal pathways for exotic or 'weedy' species (McEuen 1993). Also, Hess's (1994) modelling of disease transmission demonstrated that only under very restricted conditions would corridors lead to metapopulation extinctions. He concluded that modelled patches connected by corridors generally had fewer metapopulation extinctions than isolated patches. It also was thought that outbreeding depression (caused when increased gene flow between populations leads to loss of local variants and homogenization of gene pools) in naturally connected populations was unlikely to be a problem (Coates 1991). As few as one or two individuals moving into an area each decade could be sufficient to retain genetic diversity (Mann and Plummer 1993). Conversely, studies indicated that inbreeding depression (caused by the isolation of a population) could occur in isolated populations such as the Florida panther (*Felis concolor coryi*). Of all adult panthers known, only eight were genetically unrelated, and all adult males found in recent years have exhibited 90% infertile spermatozoa (Harris and Scheck 1991).

Other biologists, such as Hobbs and Hopkins (1991), believe that there also may be a role for corridors in a changing climate, although they acknowledge that it is difficult to determine. They propose that corridors may have the potential to allow plants and animals to make landscape-scale movements along the changing climate gradient to avoid adverse environments.

Historically, ecologists studied different elements of the landscape as separate entities. Forman (1983), Noss (1983), and others (Forman and Godron 1981; Harris 1984; Franklin and Forman 1987; Saunders and Hobbs 1991a) began to promote a different approach to the study of landscape ecology that focused on the whole landscape rather than the individual parts (Forman 1988). This new discipline depended substantially on the use of corridors and connectivity (Noss 1983; Harris and Scheck 1991).

The benefit of corridors for larger landscape areas was recognized following some of the initial use of corridors as streamside and riparian strips designed to protect fish habitat. In Australia, a number of authors proposed the use of corridors to provide a linked forest landscape (see Saunders and Hobbs 1991b). In British Columbia, the use of Forest Ecosystem Networks

Table 3.1

Potential advantages and disadvantages of corridors

Advantages	Disadvantages
1 Increase immigration rate to a reserve, which could:	1 Increase immigration rate to a reserve, which could:
(a) increase or maintain species richness and diversity (as predicted by island biogeography theory)	(a) facilitate the spread of epidemic diseases, insect pests, exotic species, weeds, and other undesirable species into reserves and across the landscape
(b) increase population sizes of particular species and decrease probability of extinction (provide a 'rescue effect') or permit re-establishment of extinct local populations	(b) decrease the genetic variation among populations or sub-populations, or disrupt local adaptations and coadapted gene complexes ('outbreeding depression')
(c) prevent inbreeding depression and maintain genetic variation within populations	
2 Provide increased foraging area for wide-ranging species	2 Facilitate spread of fire and other abiotic disturbances ('contagious catastrophes')
3 Provide predator-escape cover for movements between patches	3 Increase exposure of wildlife to hunters, poachers, and other predators
4 Provide a mix of habitats and successional stages accessible to species that require a variety of habitats for different activities or stages of their life cycles	4 Riparian strips, often recommended as corridor sites, might not enhance dispersal or survival of upland species
5 Provide alternative refugia from large disturbances (a 'fire escape')	5 Cost, and conflicts with conventional land preservation strategy to preserve endangered species habitat (when inherent quality of corridor habitat is low)
6 Provide 'greenbelts' to limit urban sprawl, abate pollution, provide recreational opportunities, and enhance scenery and land values	

Source: Noss (1987)

(FEN) was suggested as an approach to managing biodiversity across a landscape (McDougall 1991). The purpose of a FEN was to maintain natural forest ecosystem processes, to provide interior habitat, and to allow for dispersal, recolonization, and movement. These FENs would be composed of reserves, corridors, and buffers. In 1991, Noss, Soulé, and others founded the Wildlands Project, the largest landscape plan ever proposed (Mann and Plummer 1993). This project intends to establish a network of corridors, reserves, and buffers across North America to maintain continental connectivity.

Ecological Principles

Maintaining Connectivity
Connectivity is maintained in a landscape when there is a continuity of habitats and processes along environmental gradients (Noss 1991). Connectivity of ecological processes (e.g., dispersal, predation, fire, flooding) is as important as connectivity of habitats. The collective habitat requirements of species need to be viewed as interacting, functional components of the landscape ecosystem (Noss 1991). The primary intent of management should be to maintain natural connectivity at many scales, not to reconnect recently isolated habitats (Noss and Harris 1989).

Ecological processes are fundamental components of forested landscapes. Processes are characterized temporally by rates and spatially by flows. Maintaining connectivity is intended to maintain the natural continuum of these ecological rates and flows. (Photograph by Scott Harrison)

Providing habitat and movement routes are predominant themes in maintaining connectivity. The premise is that if suitable habitats are connected, fauna will move among patches to use resources, recolonize areas, and maintain gene flow (Merriam 1991b). Dispersal and gene flow are key features of these movements. Dispersal is the movement of organisms away from their place of origin; gene flow is the movement of alleles due to the dispersal of gametes of offspring (Noss 1991). Harris and Scheck (1991) present nine reasons organisms disperse:

- to forage for resources that are patchy in space, such as food, mates, or special habitats
- to exploit resources that are sporadic in time
- to exploit seasonal environments (e.g., migrations)
- to accommodate different life stages
- to return to a birthplace
- to colonize a new, local environment
- to extend range distribution
- to accommodate climate change
- to colonize new islands or continents.

They also discuss the time and distances associated with these dispersals (see Table 3 in Harris and Scheck 1991). These spatial and temporal considerations of dispersal are important. If there are barriers to long-term movements over great distances even with corridors in place, broad-scale retention of habitat, rather than corridors, may be necessary to maintain connectivity (Merriam 1991b). Nuthatches (*Sitta europaea*) in fragmented Belgian habitats had dispersal distances several times larger than nuthatches in more forested landscapes (Matthysen et al. 1995), and young nuthatches were less likely to move through the less connected landscape once they had settled. Haas (1995) found that for adult American Robins (*Turdus migratorius*), Brown Thrashers (*Toxostoma rufum*), and Loggerhead Shrikes (*Lanius ludovicianus*), movements between patches were rare but occurred significantly more frequently when sites were connected by a corridor.

Connectivity is believed to be important for viability of faunal populations (Taylor et al. 1993). The loss of connectivity, often discussed as fragmentation, is considered by some to be the greatest threat to natural biological diversity (Harris 1984; Wilcox and Murphy 1985; Wilcove et al. 1986; Noss 1991). The loss or gain of only a few species can affect many other species and destabilize the whole community (Wilson and Willis 1975). Reductions in the levels of neotropical bird populations have been shown for fragmented habitats (MacClintock et al. 1977). These migrant birds have lower dispersal and reproductive rates and exhibit specialized insectivorous feeding strategies. As connectivity is lost and habitat patches become isolated, the neotropical migrants are outcompeted by resident birds that are feeding generalists and that have all year to recolonize fragmented habitats (MacClintock et al. 1977). Populations of white-footed mice living in isolated forest patches in agricultural landscapes have lower growth rates and are more prone to extinction than populations living in forest patches linked by fencerow corridors (Fahrig and Merriam 1985).

In areas with extensive human disturbance, natural connectivity often has been lost. In Florida no natural communities are pristine (Noss and Harris 1989), and habitat management must focus on the restoration, rather than maintenance, of connectivity. This is a different situation from areas where human disturbance and habitat fragmentation are less severe. Even authors who question the cost-effectiveness of corridors as mechanisms for restoring connectivity in Florida's severely fragmented landscapes acknowledge that natural connectivity is important and should be retained before landscapes are disturbed (Simberloff and Cox 1987).

Connectivity should be maintained at the landscape level (Harris and Scheck 1991; Merriam 1991b; Noss 1991). Taylor et al. (1993) define landscape connectivity as the degree to which the landscape facilitates or impedes movement among resource patches. A landscape unit may be a mountainous watershed of 5,000 ha or an undulating plateau of 100,000 ha. In either case, the landscape unit is a distinct, measurable unit with a recognizable and repeated cluster of ecosystems and disturbance regimes (Forman and Godron 1981). Connectivity objectives should be set for each landscape unit. Objectives should be specific and focused on issues such as the viability of specific target species (Soulé 1991); rare and threatened species must receive special consideration (Bennett 1991). Csuti (1991) states that connectivity objectives should address the stabilization of natural biological diversity, not just a reduction in the rate of species loss.

Connectivity objectives need to account for all habitat disturbances within the landscape unit (Csuti 1991). The objectives must consider the duration and extent to which different disturbances will alienate habitats. Fencerows that maintain connectivity in an agricultural landscape (Merriam and Lanoue 1990) would need to be maintained indefinitely if the surrounding land remained agricultural. Conversely, linkages that maintain connectivity in a newly logged landscape could be replaced by other areas within the landscape as the new forest grows. However, Harris (1984) states that in the managed forests of the future, reserves may be 'islands' in a sea of short-rotation plantations, and that these plantations will not act as source pools to recolonize the reserves. Thus, before corridors of old-growth habitat are replaced by younger stands, there should be evidence that the new forest will provide habitat attributes similar to those of the original linkage (i.e., complex forests not structurally simplified plantations). In all cases, the

This 1996 photograph of the Bowron clearcut in central British Columbia shows how the level of connectivity in portions of this 55,000-ha clearcut has remained low 12 years after logging. (Photograph by Scott Harrison)

objectives must acknowledge that the mechanisms used to maintain connectivity will be required for decades or centuries (Csuti 1991).

Mechanisms for Maintaining Connectivity

Linkages are mechanisms by which the principles of connectivity can be achieved. Although the definitions of linkages vary, all imply that there are connections or movement among habitat patches. *Corridor* is another term commonly used to refer to a tool for maintaining connectivity. Csuti (1991) differentiates between the functions of corridors and linkages: corridors provide only pathways for movement, whereas linkages enable ecosystem processes to continue. Therefore, the successful functioning of a corridor or linkage should be judged in terms of the connectivity among subpopulations and the maintenance of potential metapopulation processes (Merriam 1991b).

Noss (1992) uses the term *linkage* to emphasize the role of faunal movements as part of the overall ecosystem function. Linkages provide routes for movement and dispersal, and habitat for non-mobile organisms; they often encompass special habitats that are distinct in the landscape, such as riparian habitats. Linkages also provide avenues for long-distance range shifts to enable species to adapt to gradual, long-term environmental changes such as global warming. Forman (1983) presents four major functions of corridors:

- to enhance the movement of flora and fauna
- to act as semi-permeable barriers to movement across the width of the corridor (the effects of water runoff, mineral-nutrient runoff, or winds can be moderated)
- to provide habitat for some species
- to provide sources of environmental and biotic effects on the surrounding matrix (organisms may move from the corridor to the surrounding matrix, colonizers may disperse, the corridor may moderate the local environment through shading or deposition of leaf litter).

The particular fauna that will benefit from linkage networks are often inferred from species' life histories, although data are limited on the extent to which linkages will be used. Bennett (1991) categorizes species by their ecological characteristics in order to speculate about their use of linkages and corridors:

Interior versus edge species. Species that can use habitat edges predominate among animals known to use corridors, but they may not need to use the corridor for movement. Species typical of habitat interiors have a greater need for corridors to minimize population isolation, but they are more limited in their use of corridors. Predation along habitat edges is believed to be an important influence on interior species and their ability to use corridors effectively.

Habitat specialists versus generalists. Because of specialized habitat requirements, certain species occur in low densities, and their conservation status would be enhanced by corridors that effectively link populations. However, specialized habitats may not be available or are difficult to incorporate and manage in corridors, thus limiting the value of the corridor to these species. In contrast, species with generalized habitat requirements are more likely to use corridors, but because of their wider habitat tolerance they may not need to do so.

Body size and scale of movement. Large species and those capable of longer regular movements are more likely to be able to use corridors for direct movements by individuals. They may also be less vulnerable to predation within the corridor than are small animals and those that move shorter distances per unit time. However, smaller and less mobile animals may maintain a resident population within the corridor, which in turn serves as an additional source of dispersing individuals.

In general, species that need to use corridors are those that have an obligate relationship with natural vegetation and are dependent on the remnant system of habitat (Bennett 1991). Moreover, even relatively little use of corridors may be enough. In a population model examining the status of the endangered Florida panther, one or two dispersals every decade was enough to keep the population from going extinct (Mann and Plummer

1993). MacClintock et al. (1977) felt that most forest species would sustain breeding populations in larger fragments of habitat (thousands of acres) provided the fragments were linked to larger forest tracts.

Despite gaps in our understanding of the mechanisms for maintaining connectivity within landscapes, it is better to retain corridors and then assess their role than to lose corridors and find out they were important (Saunders and Hobbs 1991a). Bennett (1991) summarizes the situation as follows:

> Despite our limited knowledge concerning the use of corridors by plants and animals, we must proceed with the designation and management of corridors on the basis that they *are* beneficial to nature conservation. Until further knowledge is available, management should endeavour to: retain existing corridors and corridor networks; maintain corridors as close to a 'natural' or 'pristine' state as is possible, in order to leave future management options open; and avoid activities or disturbance that will reduce corridor width or lead to an increase in 'edge' effects in the corridor.

Corridor Location and Design
Corridors are seldom thought of as single strips of natural vegetation but rather as an integral part of an interconnected landscape. The specific purpose of a corridor will influence its location (Harrison 1992). The location of corridors is important because corridors attempt to accommodate so many aspects of faunal ecology: species movement type and magnitude, immigration and emigration rates, movement rates, habitat requirements, and interactions among organisms (Soulé 1991).

Corridor networks should link ridgetops and valley bottoms and encompass a diversity of habitats and topographic gradients (Noss and Harris 1989; Noss 1991). Lindenmayer and Nix (1993) found that corridors that linked the valley bottom to the ridgetop had more species use than corridors located only along a midslope. Similarly, riparian habitats are often candidates as corridors because of the ecological importance of these areas and because the dendritic drainage patterns usually link midslopes and valley bottoms (Harris 1984). However, the maintenance of riparian corridors alone may not be suitable for some upland species (Noss 1991). In general, corridor location and design should reflect the ecology of an area. Corridors should be designed so that they remain functional after a human-caused or natural disturbance and until the surrounding habitat matrix can return to a pre-disturbance state.

Different organisms have different dispersal capacities (e.g., birds and bats versus terrestrial snails and plants), and a movement corridor for one species may be a barrier to another (Noss 1991). Information on the requirements for the full gamut of native species is therefore required to design

In contrast to more typical alpine mountain goat (*Oreamnos americanus*) populations, some mountain goats inhabit river canyons surrounded by forested habitat. For these canyon-dwelling goats, it is unclear whether the forest acts as a corridor or a barrier. (Photograph by Georgie Harrison)

and evaluate the effectiveness of corridors on the basis of their dispersal performance (Harris and Scheck 1991). In most cases, corridors intended as pathways for movement should be continuous for maximum effectiveness (Forman and Godron 1981). Soulé (1991) believes that they should have straight sides and a constant width because animals will spend less time in edge habitats. He argues that any departure from linearity (e.g., corridors with barriers or doglegs, or shaped as cul-de-sacs) will create a maze effect that can trap animals. McEuen (1993), however, argues that if the habitat within the corridor is sufficient 'survival' habitat, being caught in this corridor is the same as being caught in a reserve.

The quality of the habitat within corridor networks is an important element of connectivity (Henein and Merriam 1990). Frequently, corridors are intended to facilitate the movement of organisms by extending interior forest habitat characteristics through areas of timber harvest (Saunders and Hobbs 1991a). Although planning decisions can be eased by simply placing corridors in 'inoperable' or 'non-productive' areas where timber harvest is not feasible, maintaining connectivity among 'productive' timber areas is probably necessary for species that prefer such habitats. Harris and Scheck (1991) emphasize that faunal dispersal corridors should maintain or enhance the habitat values for native fauna and flora; corridors should therefore consist of natural or restored native landscape features. Harris and Scheck

(1991) cite an example where introduced landscape features have resulted in the loss of native biological diversity: trees planted as windbreaks on the Prairies provided an unnatural corridor; Yellow-Shafted Flickers (*Colaptes auratus*) followed these corridors westward and interbred with the western Red-Shafted Flickers, resulting in the loss of the subspecies variation.

Corridor Width

The suitability of a corridor for a particular species may depend on behaviour, physical aspects of the landscape, or distance between patches. Scale also influences the definition of a corridor; for example, fencerows can provide linkages for small mammals but not for grizzly bears (*Ursus arctos*) or cougars. Saunders and de Rebeira (1991) found that birds used vegetated connections as narrow as 4 m for movement; however, wide-ranging, large mammals require landscape-scale corridors (Noss 1991). Some authors suggest that corridor width must be established in relation to the movement ability of a species, otherwise the corridor may act as a sink habitat rather than a channel (Saunders and Hobbs 1991a; Soulé 1991). Csuti (1991) points out that the concerns about wide corridors and 'sinks' is based on a view of corridors as strictly travel routes. He notes that if a corridor is wide enough, some species will maintain a population within it and will eventually die natural deaths within the corridor.

Ultimately, the effectiveness of a corridor depends on the amount of human activity in and around the corridor (Harrison 1992), and the optimum corridor width depends on the strength of edge effect (Soulé 1991). Forman and Godron (1981) outline the importance of corridors as buffers against edge effects:

> The corridor must provide protective cover for species from natural predators, domestic animals, and human effects lining each side of the corridor. The outer portions of the strip corridor have the edge effect, while the central portion contains the interior environment required for many patch interior species. For this reason, the width of a strip corridor is critical, since the interior environment must be present and sufficiently wide itself to be used by interior species.

A range of edge effects occur along the interface of interior and disturbed habitat types: (1) biological edge effects (predation, herbivory, weed invasion), (2) physical effects (light, temperature, humidity, nutrient status, hydrology changes, wind effects), and (3) human effects (pollution, hunting) (Start 1991). Importantly, the greater the difference between the corridor and the surrounding habitat matrix, the deeper the edge effect (Forman 1983; Harris 1984). Feathering edges around reserves and corridors (e.g., partial-cut logging instead of clearcutting with hard edges) is one method of mitigating some edge effects. Other authors feel that the solution is more straightforward: to minimize the edge effect, widen the corridor (Start 1991).

A single optimum width is impossible to specify. Corridor width should be assessed in relation to the life histories and requirements of the fauna for which the corridors are intended (Friend 1991). In Maine, species known to do well in edge habitats – moose (*Alces alces*), white-tailed deer (*Odocoileus virginianus*), and black bear (*Ursus americanus*) – used corridors 76-152 m wide (Laitin 1987). Csuti (1991) uses measures of edge effects to present minimum useful corridor widths, and suggests widths of 1.2-6.4 km to provide interior habitat within the corridor. Harris and Scheck (1991) account for spatial and temporal aspects of corridor design (capitals their emphasis):

WHEN THE MOVEMENT OF INDIVIDUAL ANIMALS IS BEING CONSIDERED, WHEN MUCH IS KNOWN ABOUT THEIR BEHAVIOUR, AND WHEN THE CORRIDOR IS EXPECTED TO FUNCTION IN TERMS OF WEEKS OR MONTHS THEN THE APPROPRIATE CORRIDOR WIDTH CAN BE MEASURED IN METRES (*c.* 1-10 m).

WHEN THE MOVEMENT OF A SPECIES IS BEING CONSIDERED, WHEN MUCH IS KNOWN ABOUT ITS BIOLOGY, AND WHEN THE CORRIDOR IS TO FUNCTION IN TERMS OF YEARS THEN THE CORRIDOR WIDTH SHOULD BE IN 100s OF METRES (*c.* 100-1000 m).

WHEN THE MOVEMENT OF ENTIRE ASSEMBLAGES OF SPECIES IS BEING CONSIDERED AND/OR WHEN LITTLE IS KNOWN OF THE BIOLOGY OF THE SPECIES AND/OR IF THE FAUNAL DISPERSAL CORRIDOR IS TO FUNCTION OVER DECADES THE APPROPRIATE WIDTH MUST BE MEASURED IN KILOMETRES.

There are few data indicating how corridor width and composition may influence faunal movement, although many authors feel that these are important considerations with respect to the effectiveness of a corridor network (Friend 1991). In broad terms, for a faunal dispersal corridor to be effective, its size must be appropriate for the ecological function being asked of it; the width must be appropriate to the scale of the phenomenon being addressed (Harris and Scheck 1991). Bigger animals probably need wider corridors, and corridor networks in heavily disturbed ecosystems must provide more habitat for connectivity of ecological functions. Of course, all corridors, regardless of their width, must connect areas of useful, protected habitat (Harris and Scheck 1991).

Summary

Connectivity connotes connections among habitats, species, communities, and ecological processes. Connectivity is the antithesis of fragmentation. Corridors are tools by which the principle of connectivity is maintained. Despite some potential disadvantages, corridors are the only means of maintaining connectivity between patches for mobile organisms (Csuti 1991). One should accept the working hypothesis that wide, continuous corridors are most likely to foster movements (Saunders and Hobbs 1991a).

Research Needs

Maintaining connectivity throughout the landscape is a relatively recent conservation concern. The following are the major information gaps identified in the literature.

Monitoring of corridor use. In order to understand the success or failure of connectivity through corridors, it is imperative to monitor the species that use them (Noss 1987; Saunders and Hobbs 1991a; McEuen 1993). As well, monitoring the shift of species that use corridors when conditions change (e.g., width, length, habitat) could be significant to future corridor design (McEuen 1993). For experimental designs and statistical considerations, see Nicholls and Margules (1991) and Inglis and Underwood (1992).

Genetic studies. It is important to learn more about the effects of inbreeding and outbreeding on both target and corridor-dependent species. The effectiveness of corridors in maintaining genetic diversity and gene flow also needs to be understood (Coates 1991).

Fragmentation. Studies should be initiated on how fragmentation affects the persistence of amphibian, reptile, and mammal species during the process of fragmentation, as opposed to studies on the after-effects of fragmentation (i.e., habitat 'islands') (Verner 1984; Lehmkuhl and Ruggiero 1991). There is also a need to determine population parameters and then model population responses to fragmentation over time and space (Lehmkuhl and Ruggiero 1991).

Behaviour. Information such as behaviour, habitat requirements, and life histories is needed on 'keystone' species (large mammals, endangered species, or corridor-dependent species) to better understand their needs for connectivity (Friend 1991; Merriam 1991a).

Width and dimensions of corridors. Studies to determine the importance of width or width versus length of a corridor should be initiated (Friend 1991). Also, studies of 'keystone' species and their home ranges may help determine the minimum widths required for a corridor (Harrison 1992). If corridors are to support the movement of species over a long period, it will be important to determine the optimum dimensions needed to support resident populations of a wide range of native species (Bennett 1990).

Dispersal. Studying dispersal in natural corridors and observing species movement across unknown and unsuitable habitat may help in developing future landscape designs for connectivity (Harrison 1992). Data such as distance, rate, and frequency of species movement through corridors and the landscape should be collected (Bennett 1991).

Invertebrate studies. Understanding invertebrate assemblages in corridors may be key to corridor design, as invertebrates are an integral part of an ecosystem and an important food source for many vertebrates (Saunders and Hobbs 1991a).

Literature Cited

Beier, P. 1993. Determining minimum habitat areas and habitat corridors for cougars. Conservation Biology 7(1):94-108.

Bennett, A.F. 1990. Habitat corridors and the conservation of small mammals in a fragmented forest environment. Landscape Ecology 4(2-3):109-122.

–. 1991. What types of organisms will use corridors. Pages 407-408 in D.A. Saunders and R.J. Hobbs (eds.). Nature conservation 2: the role of corridors. Surrey Beatty and Sons, NSW, Australia. 442 pp.

Cable, T.T. 1991. Windbreaks, wildlife, and hunters. Pages 36-55 in J.E. Rodiek and E.G. Bolen (eds.). Wildlife and habitats in managed landscapes. Island Press, Washington, D.C. 219 pp.

Coates, D.J. 1991. Gene flow along corridors. Pages 408-409 in D.A. Saunders and R.J. Hobbs (eds.). Nature conservation 2: the role of corridors. Surrey Beatty and Sons, NSW, Australia. 442 pp.

Csuti, B. 1991. Introduction. Pages 81-90 in W.E. Hudson (ed.). Landscape linkages and biodiversity. Defenders of Wildlife, Washington, D.C. 196 pp.

Date, E.M., H.A. Ford, and H.F. Recher. 1991. Frugivorous pigeons, stepping stones, and weeds in northern New South Wales. Pages 241-245 in D.A. Saunders and R.J. Hobbs (eds.). Nature conservation 2: the role of corridors. Surrey Beatty and Sons, NSW, Australia. 442 pp.

Diamond, J.M. 1975. The island dilemma: lessons of modern biogeography and the design of nature reserves. Biological Conservation 7:129-146.

Fahrig, L., and G. Merriam. 1985. Habitat patch connectivity and population survival. Ecology 66(6):1762-1768.

Forman, R.T.T. 1983. Corridors in a landscape: their ecological structure and function. Ekologia 2(4):375-387.

–. 1988. Landscape ecology plans for managing forests. Pages 131-135 in Proceedings of the Convention for the Society of American Foresters. USDA Forest Service General Technical Report NE-140.

–. 1991. Landscape corridors: from theoretical foundations to public policy. Pages 71-84 in D.A. Saunders and R.J. Hobbs (eds.). Nature conservation 2: the role of corridors. Surrey Beatty and Sons, NSW, Australia. 442 pp.

Forman, R.T.T., and M. Godron. 1981. Patches and structural components for a landscape ecology. BioScience 31:733-739.

Franklin, J.F., and R.T.T. Forman. 1987. Creating landscape patterns by forest cutting: ecological consequences and principles. Landscape Ecology 1(1):5-18.

Friend, G.R. 1991. Does corridor width or composition affect movement? Pages 404-405 in D.A. Saunders and R.J. Hobbs (eds.). Nature conservation 2: the role of corridors. Surrey Beatty and Sons, NSW, Australia. 442 pp.

Gates, J.E. 1991. Powerline corridors, edge effects, and wildlife in forested landscapes of the Central Appalachians. Pages 14-32 in J.E. Rodiek and E.G. Bolen (eds.). Wildlife and habitats in managed landscapes. Island Press, Washington, D.C. 219 pp.

Harris, L.D. 1984. The fragmented forest: island biogeography theory and the preservation of biotic diversity. University of Chicago Press, Chicago, Ill. 211 pp.

Harris, L.D., and K. Atkins. 1991. Faunal movement corridors in Florida. Pages 117-134 in W.E. Hudson (ed.). Landscape linkages and biodiversity. Defenders of Wildlife, Washington, D.C. 196 pp.

Harris, L.D., and J. Scheck. 1991. From implications to applications: the dispersal corridor principle applied to the conservation of biological diversity. Pages 189-220 in D.A. Saunders and R.J. Hobbs (eds.). Nature conservation 2: the role of corridors. Surrey Beatty and Sons, NSW, Australia. 442 pp.

Harrison, R.L. 1992. Toward a theory of inter-refuge corridor design. Conservation Biology 6(2):293-295.

Harrison, S.P. 1991. Local extinction in a metapopulation context: an empirical evaluation. Biological Journal of the Linnean Society. 42:73-88.

Harrison, S., B. Chatterson, and D. Paul. In press. Landscape connectivity and its effect on the ecological processes of spruce grouse populations. Wildlife Biology.

Hass, C.A. 1996. Dispersal and use of corridors by birds in wooded patches on an agricultural landscape. Conservation Biology 9(4):845-854.

Hay, K.G. 1991. Greenways and biodiversity. Pages 162-175 in W.E. Hudson (ed.). Landscape linkages and biodiversity. Defenders of Wildlife, Washington, D.C. 196 pp.

Henein, K., and G. Merriam. 1990. The elements of connectivity where corridor quality is variable. Landscape Ecology 4(2/3):157-170.

Hess, G.R. 1994. Conservation corridors and contagious disease: a cautionary note. Conservation Biology 8(1):256-262.

Hobbs, R.J. 1992. The role of corridors in conservation: solution or bandwagon? Tree 7(11):389-392.

Hobbs, R.J., and A.J.M. Hopkins. 1991. The role of conservation corridors in a changing climate. Pages 281-290 in D.A. Saunders and R.J. Hobbs (eds.). Nature conservation 2: the role of corridors. Surrey Beatty and Sons, NSW, Australia. 442 pp.

Inglis, G., and A.J. Underwood. 1992. Comments on some designs proposed for experiments on the biological importance of corridors. Conservation Biology 6(4):581-586.

Kucera, T.E., and C. McCarthy. 1988. Habitat fragmentation and mule deer migration corridors: a need for evaluation. Transactions of the Western Section of the Wildlife Society 24:61-67.

Laitin, J. 1987. Corridors for wildlife. American Forests (September/October):47-49.

Lehmkuhl, J.F., and L.F. Ruggiero. 1991. Forest fragmentation in the Pacific Northwest and its potential effects on wildlife. Pages 35-46 in L.F. Ruggiero, K.B. Aubry, A.B. Carey, and M.H. Huff (eds.). Wildlife and vegetation of unmanaged Douglas-fir forests, USDA Forest Service General Technical Report GTR-285.

Levins, R. 1970. Extinction. Pages 77-107 in M. Gerstenhaber (ed.). Some mathematical questions in biology: lectures on mathematics in the life sciences, vol. 2. American Mathematical Society, Providence, R.I.

Lindenmayer, D.B., and H.A. Nix. 1993. Ecological principles for the design of wildlife corridors. Conservation Biology 7(3):627-630.

MacArthur, R.H., and E.O. Wilson. 1967. The theory of island biogeography. Princeton University Press, Princeton, N.J. 442 pp.

MacClintock, L., R.F. Whitcomb, and B.L. Whitcomb. 1977. Island biogeography and 'habitat islands' of eastern forest. American Birds 31(1):3-16.

Mann, C.C., and M.L. Plummer. 1993. The high cost of biodiversity. Science 260:1868-1871.

Matthysen, E., F. Adriaensen, and A.A. Dhondt. 1995. Dispersal distances of nuthatches, *Sitta europaea*, in a highly fragmented forest habitat. Oikos 72:375-381.

McDougall, I.A. 1991. Forest ecosystem networks – a strategy for managing biological diversity. Unpublished report, B.C. Ministry of Environment, Lands and Parks. 7 pp.

McEuen, A. 1993. The wildlife corridor controversy: a review. Endangered Species Update 10(11/12):1-6/12.

Merriam, G. 1991a. Corridors and connectivity: animal populations in heterogeneous environments. Pages 133-142 in D.A. Saunders and R.J. Hobbs (eds.). Nature conservation 2: the role of corridors. Surrey Beatty and Sons, NSW, Australia. 442 pp.

–. 1991b. Are corridors necessary for the movement of biota? Pages 406-407 in D.A. Saunders and R.J. Hobbs (eds.). Nature conservation 2: the role of corridors. Surrey Beatty and Sons, NSW, Australia. 442 pp.

Merriam, G., and A. Lanoue. 1990. Corridor use by small mammals: field measurement for three experimental types of *Peromyscus leucopus*. Landscape Ecology 4(2/3):123-131.

Naiman, R.J., H. Decamps, and M. Pollock. 1993. The role of riparian corridors in maintaining regional biodiversity. Ecological Applications 3(2):209-212.

Newmark, W.D. 1987. A land-bridge island perspective on mammalian extinctions in western North American parks. Nature 325:430-432.

Nicholls, A.O., and C.R. Margules. 1991. The design of studies to demonstrate the biological importance of corridors. Pages 49-61 in D.A. Saunders and R.J. Hobbs (eds.). Nature conservation 2: the role of corridors. Surrey Beatty and Sons, NSW, Australia. 442 pp.

Noss, R.F. 1983. A regional landscape approach to maintain diversity. BioScience 33(11):700-706.

–. 1987. Corridors in real landscapes: a reply to Simberloff and Cox. Conservation Biology 1(2):159-164.

–. 1991. Landscape connectivity: different functions at different scales. Pages 27-39 in W.E. Hudson (ed.). Landscape linkages and biodiversity. Defenders of Wildlife, Washington, D.C. 196 pp.

–. 1992. North American wilderness recovery project: land conservation strategy. Wild Earth special issue. 35 pp.

Noss, R.F., and L.D. Harris. 1989. Habitat connectivity and the conservation of biological diversity: Florida as a case history. Pages 131-135 in Forestry of the Frontier Convention of the Society of American Foresters, Spokane, Wash. 444 pp.

Preston, F.W. 1962. The canonical distribution of commonness and rarity. Part 2. Ecology 43(3):410-432.

Saunders, D.A., and C.P. de Rebeira. 1991. Values of corridors to avian populations in a fragmented landscape. Pages 221-240 in D.A. Saunders and R.J. Hobbs (eds.). Nature conservation 2: the role of corridors. Surrey Beatty and Sons, NSW, Australia. 442 pp.

Saunders, D.A., and R.J. Hobbs. 1991a. The role of corridors in conservation: what do we know and where do we go? Pages 421-427 in D.A. Saunders and R.J. Hobbs (eds.). Nature conservation 2: the role of corridors. Surrey Beatty and Sons, NSW, Australia. 442 pp.

– (eds.). 1991b. Nature conservation 2: the role of corridors. Surrey Beatty and Sons, NSW, Australia. 442 pp.

Simberloff, D., and L.G. Abele. 1976. Island biogeography theory and conservation practice. Science 191:285-286.

Simberloff, D., and J. Cox. 1987. Consequences and costs of conservation corridors. Conservation Biology 1(1):63-71.

Simberloff, D., J.A. Farr, J. Cox, and D.W. Mehlman. 1992. Movement corridors: conservation bargains or poor investments? Conservation Biology 6(4):493-504.

Simpson, G.G. 1936. Data on the relationship of local and continental mammalian fauna. Journal of Palaeontology 10:410-414.

–. 1940. Mammals and land bridges. Journal of the Washington Academy of Sciences 30:137-163.

Soulé, M.E. 1991. Theory and strategy. Pages 91-103 in W. Hudson (ed.). Landscape linkages and biodiversity. Defenders of Wildlife, Washington, D.C. 196 pp.

Start, A.N. 1991. How can edge effects be minimized? Pages 417-418 in D.A. Saunders and R.J. Hobbs (eds.). Nature conservation 2: the role of corridors. Surrey Beatty and Sons, NSW, Australia. 442 pp.

Szacki, J. 1987. Ecological corridor as a factor determining the structure and organization of a bank vole population. Acta Theriologica 32(3):31-44.

Taylor, P.D., L. Fahrig, K. Henein, and G. Merriam. 1993. Connectivity is a vital element of landscape structure. Oikos 68(3):571-573.

Terborgh, J. 1974. Preservation of natural diversity: the problem of extinction prone species. BioScience 24:715-722.

Verner, J. 1984. Predicting effects of habitat patchiness and fragmentation – the researcher's viewpoint. Pages 327-329 in J. Verner, M.L. Morrison, and C.J. Ralph (eds.). Wildlife 2000: modeling habitat relationships of terrestrial vertebrates. University of Wisconsin Press, Madison, Wis.

Wilcove, D.S., C.H. McLennan, and A.P. Dobson. 1986. Habitat fragmentation in the temperate zone. Pages 237-256 in M.E. Soulé (ed.). Conservation biology: the science of scarcity and diversity. Sinauer Associates, Sunderland, Mass.

Wilcox, B.A., and D.D. Murphy. 1985. Conservation strategy: the effects of fragmentation on extinction. American Naturalist 125:879-887.

Wilson, E.O., and E.O. Willis. 1975. Applied biogeography. Pages 522-534 in M. Cody and J.M. Diamond (eds.). Ecology and evolution of communities. Harvard University Press, Cambridge, Mass.

4
Riparian Areas and Wetlands
Joan Voller

Definitions

Riparian areas: There is no standard definition of a riparian area (Melton et al. 1983). The word *riparian* is derived from the Latin for bank or shore, and simply refers to land adjacent to a body of water. However, depending on the purpose and the user of the definition, its meaning can become more complex. Definitions may take into consideration soil moisture conditions, vegetation, or geomorphology. A natural riparian ecosystem will therefore vary in width and shape along its length as these conditions vary in width and shape (Gregory and Ashkenas 1990). A riparian area is also considered to be three-dimensional by extending outward from the high-water mark as well as upward into the canopy (Gregory and Ashkenas 1990). The following two definitions describe a natural riparian ecosystem.

The Willamette National Forest's definition of riparian area is: 'The aquatic ecosystem and the portions of the adjacent terrestrial ecosystem that directly affect or are affected by the aquatic environment. This includes streams, rivers, and lakes and their adjacent side channels, floodplains and wetlands. The riparian area includes portions of hillslope that serve as streamside habitats for wildlife' (Gregory and Ashkenas 1990).

The definition in the British Columbia Forest Practices Code is: 'The land adjacent to the normal high water line in a stream, river, or lake, extending to the portion of land that is influenced by the presence of the adjacent ponded or channeled water. Riparian areas typically exemplify a rich and diverse vegetative mosaic reflecting the influence of available surface water' (B.C. Ministry of Forests 1994).

The Willamette National Forest and the B.C. Forest Practices Code also define a Riparian Management Zone (RMZ) or Riparian Management Area (RMA). Both definitions refer to an arbitrary line drawn along the side of the riparian area under which certain management applications will take place, usually, or preferably, on a site-specific basis.

Riparian Management Zone: 'site-specific boundaries established to meet riparian management objectives in riparian areas. Riparian management zones are contained within but do not necessarily include the entire riparian area' (Gregory and Ashkenas 1990).

Riparian Management Area: 'a classified area of specified width surrounding or adjacent to streams, lakes, riparian areas, and wetlands. The RMA includes, in many cases, adjacent upland areas. It extends from the top of the stream bank (bank full height) or from the edge of a riparian area or wetland or the natural boundary of a lake outward to the greater of: 1) the specified RMA distance, 2) the top of the inner gorge, or 3) the edge of the flood plain. Where a riparian area or wetland occurs adjacent to a stream or lake, the RMA is measured from the outer edge of the wetland' (B.C. Ministry of Forests 1994).

Wetlands: Unlike riparian areas, wetlands have been defined more concisely. A wetland is defined as 'land that is saturated with water long enough to promote wetland or aquatic processes as indicated by poorly drained soils, hydrophytic vegetation, and various kinds of biological activity which are adapted to a wet environment' (National Wetlands Working Group 1993, in North American Wetlands Conservation Council [Canada] 1993).

In addition, the B.C. Forest Practices Code defines a wetland as follows:

> Those land forms, as distinguished on 1:20 000 air photos, wet enough and inundated frequently enough to develop and support natural vegetation cover that is distinct from the vegetation in the neighbouring, freely drained upland sites. These include but are not limited to swamps, marshes, bogs, and similar areas.
>
> The outer edge of the wetland is considered to be the point at which there is an obvious shift to upland forest from the grasses, shrubs, deciduous trees, and coniferous trees that characterize the wetland (B.C. Ministry of Forests 1994).

Kistritz and Porter (1993) defined seven classes of wetlands for British Columbia:

Bog: a peatland with the water table at or near the surface. The bog surface may be raised or at the same level as the surrounding terrain. It is generally acidic and low in nutrients because it is essentially unaffected by the surrounding nutrient-rich runoff. Precipitation is the main source of water. A bog has an organic substrate made mostly of sphagnum peat. Trees may or may not be present but bogs are usually covered with species of sphagnum and ericaceous shrubs.

Fen: a peatland with the water table just above or below the surface. Mineral soils make the water nutrient-rich and minerotrophic. Fen water is not

An example of a marsh found along the Cowichan River, Vancouver Island. (Photograph by Joan Voller)

stagnant because of constant seepage or channellization of water through the peat. A fen has an organic substrate made mostly of sedge and/or brown moss peat. The vegetation usually consists of sedges or shrubs dominated by brown mosses.

Marsh: a mineral wetland that has a seasonal or permanent water depth up to 2 m. The water may be either standing or slowly moving and is nutrient-rich, ranging from fresh to saline. The substrate is mostly mineral. The vegetation (sedges, grasses, rushes, and reeds) is non-woody and grows in banded patterns

Swamp: a wooded, mineral wetland with nearly permanent, gently flowing water. The water is rich in minerals and nutrients. The substrate is woody peat or mineralized material. The vegetation is dominated by flood-tolerant deciduous and coniferous trees or tall shrubs, herbs, and mosses.

Wet meadow: a mineral wetland that is seasonally saturated but seldom submerged. The water table drops below the rooting level for most of the growing season. A wet meadow has mineral-rich soils. There is herbaceous vegetation with a grassy appearance, mostly flood-tolerant grasses, low sedges, rushes, and forbs.

Shrub carr: a shrub-dominated wetland that is seasonally saturated but seldom submerged. The water table drops below the rooting level during the growing season. The substrate is made up of mineral soils with a thin top layer of organics up to 15 cm thick. The shrubs may grow between 1 and 3 m in height and include scrub birch and willows.

Shallow water: open waters that are intermittently or permanently flooded, with midsummer depths of less than 2 m. Shallow water has a summer open water zone of greater than 75% of the wetland surface area or an area greater than 8 ha. These open expanses may be referred to as ponds, pools, shallow lakes, oxbows, reaches, channels, or impoundments. Vegetation consists of submerged or floating aquatics.

Further components of wetland classification describe vegetation physiognomy (structure), site series, hydrology, hydrochemistry, substrate, and hydrogeomorphic landscape characteristics (Kistritz and Porter 1993; B.C. Wetlands Working Group, in prep.).

Background

Timber harvesting began on the west coast of North America in the mid to late 1700s. By the late 1800s, most of the timber on the coast and along large rivers had been harvested in the U.S. Pacific Northwest and lower coastal B.C. Soon after, harvest of the interior forests began. Streams and rivers became the transportation corridors for moving the harvested logs to the mills. These waterways were cleared of coarse woody debris, and splash dams were built on many to ease the transport of logs downriver. The surges of water and moving logs soon scoured the pools and stream banks, destroying fish habitat in the process. In some cases, the dams desiccated areas, causing the death of eggs, alevins, and fry (Meehan 1991).

Until the public outcry began in the 1960s, commodity production took precedence over environmental concerns. Timber harvest was undertaken with little concern for fish and wildlife habitat (Hartman and Scrivener 1990; Meehan 1991). Over the last 200 years, more than 80% of natural riparian areas have disappeared in North America and Europe (Naiman et al. 1993). Benke (1990) found that only 2% of 3.25 million stream miles surveyed in the U.S. (excluding Alaska) were considered of 'high natural quality.' As well, Dahl (1990) found that approximately 53% of all wetlands in the U.S. (excluding Alaska) had been lost. Only 9% of California's wetlands remain (FEMAT 1993). Horwitz (1978) stated that more than 'two thirds of all commercial fish and shellfish species in the Gulf, Atlantic, and Pacific fisheries in U.S.A. waters rely on (coastal) wetlands for shelter or food.' Wilcove et al. (1992) stated that habitat loss was one of the leading causes of native fish extinctions in North America.

Approximately 6% of British Columbia's total land area is covered by wetlands (Van Ryswyk et al. 1992). Loss and conversion of these wetlands has been severe: for example, approximately 70% of natural Pacific estuarine marshes (Sustaining Wetlands Forum 1990), 76% of the wetlands around Victoria, and 85-90% of the area of large marshes between Penticton and Osoyoos, B.C., have been lost (Lang 1991). Even as recently as 1987, the Canadian Forest Service encouraged the drainage of wetlands to 'improve' them for forestry (Hillman 1987). A significant amount of wetlands in B.C. are also used for forage and horticultural crops: half the beef cattle of the Cariboo-Chilcotin are fed from forage produced on wetlands in this area (Van Ryswyk et al. 1992). Recently, however, concern for conservation of wetlands has been increasing. In a symposium on wetlands and peatlands, Pojar and Roemer (1987) expressed concern that the most significant wetlands in B.C. (the lower Fraser Valley and eastern Vancouver Island) were already badly disturbed or beyond repair.

Even though research on the components of wetlands and riparian areas has been going on since the early 1900s, only recently have these areas received attention as important components of the ecosystem. For example, their classification is a recent phenomenon. The idea of a comprehensive Canadian wetland classification system was first envisaged in 1973, and in 1975 Zoltai et al. proposed a hierarchical, ecologically based system (North American Wetlands Conservation Council [Canada] 1993). In 1980 Tarnocai refined the system to better suit Canadian wetlands. In 1981 Runka and Lewis developed a wetland classification system for the Cariboo-Chilcotin area of British Columbia. In 1987 the first edition of the 'Canadian Wetland Classification System' was developed from a blend of existing systems. This system was further revised in 1993 by the National Wetlands Working Group (North American Wetlands Conservation Council [Canada] 1993). Revisions are currently being made to B.C.'s wetland classification system by the B.C. Wetland Working Group and the Ecosystem Working Group of the Resource Inventory Committee (A. Banner, pers. comm., 1996). Riparian areas surrounding lakes, rivers, and streams still do not have a universal classification system; this is also being addressed by the B.C. Wetland Working Group. However, classification systems exist for various components of the riparian area, such as an ecosystem classification system (Meidinger and Pojar 1991) and a soil classification system (Agriculture Canada Expert Committee and Soil Survey 1987).

Not until the mid-1950s was the need to integrate B.C.'s fisheries and forestry resources recognized and a referral system established (Hartman and Scrivener 1990). However, the referral system was based on information from studies made in areas very unlike coastal B.C. In response to this lack of local knowledge, the Canadian Department of Fisheries and Oceans initiated the Carnation Creek project on Vancouver Island in 1970 (Hartman and Scrivener 1990). This project was designed to examine the effects of forest harvest activities on water, soil, vegetation, and fish. The study continues today but is now administered by the B.C. Ministry of Forests, the B.C. Ministry of Environment, Lands and Parks, and Forestry Canada.

Since the 1970s many studies have looked at the effects of logging on fisheries resources. Many studies carried out in the 1970s and 1980s indicated that primary production, and in some cases salmonid biomass, increased in streams flowing through clearcuts. Studies by Oregon State University researchers indicated that primary production increased 50-100% in clearcut versus forested areas (Gregory 1976).

However, Carnation Creek and other studies demonstrated that changes in streamside habitat were associated with a shift in the relative abundance and/or age structure of species using the stream. Bisson and Sedell (1982) found that average salmonid biomass was two times greater in clearcut streams, but this increase was accompanied by shifts in species and age composition, thus decreasing the relative abundance of species such as coho (*Oncorhynchus kisutch*) from the area. Chum salmon (*Oncorhynchus keta*) production in Carnation Creek also declined sharply after logging. Further studies indicated that because of the interdependence of species within a stream, the effect of a temperature change on one organism could affect the whole

Mature forest stands on the edge of a stream provide steady input of large organic debris for bank stabilization and bedload retention and provision of in-stream fish habitat. (Photograph by Steve Voller)

food web (Miller 1987). A study in Alaska by Myren and Ellis (1982) found that clearcutting to the edge of streams could cause drastic changes in stream flows over time: evapotranspiration in growing second growth could drastically reduce stream flow in the summer and thus affect the spawning success of species such as pink (*Oncorhynchus gorbuscha*) and chum salmon and limit habitat for coho and invertebrates.

Studies since the 1970s have also indicated that increased sediment mobilization may be detrimental to invertebrates (Miller 1987) and a main factor in lowering the quality of fish habitat (Hollingsworth 1988). Studies have indicated the wide-ranging hazards of increased sediment loads in streams. Sediment has been found to fill in pools, contribute to changes in stream morphology, as well as bury or suffocate eggs by filling in spawning gravel, block gills of fry, reduce visual feeding efficiencies in fish, and reduce streambed oxygen levels (Corbett et al. 1978; Gregory and Ashkenas 1990).

Research on the effectiveness of forested buffer strips as filters is relatively limited (Belt et al. 1992). One study indicated that stream bank erosion increased 250% after clearcutting, but if vegetated buffer strips were used, erosion increased only 32% (Belt et al. 1992). The role of large organic debris (in-stream coarse woody debris) in creating and maintaining stream channel structure and sufficient aquatic habitat, and in controlling sediment deposition, has been extensively studied over the last few years (Sedell et al. 1988) (for more information on large organic debris, refer to Chapter 7).

Like the results of studies on fish populations, invertebrate species composition was found to shift with a change in streamside vegetation (Triska et al. 1982). Because invertebrates are a large part of salmonid diets, interest in understanding invertebrate life cycles has increased over the years. Jackson and Resh (1989) found that most aquatic insects spent the reproductive part of their life cycles as flying adults in the riparian area, and that changes in the abiotic or biotic environment of the riparian area could directly affect their reproductive success. Cannings (1990) pointed out how little we know about the invertebrate fauna of B.C.: a collecting trip to the west coast of Vancouver Island yielded 519 species, of which 31 were undescribed and a further 34 species were unknown to Canada.

Like studies on fisheries resources, comprehensive research into the importance of wetlands and riparian areas for wildlife did not begin in earnest until the 1970s (Meehan 1991). Riparian habitats are thought to be one of the most important habitat types from a multispecies perspective, supporting more species, as well as higher population densities, than upland habitat (Melton et al. 1983). Stevens et al. (1977) found twice as many breeding birds in riparian compared with non-riparian plots. Bull (1978) found that cavity-nesters preferred riparian habitats and some were found only in this habitat. Research over several years has demonstrated the existence of riparian-dependent birds, mammals, and amphibians (Oakley et al. 1985). Cross (1985) found that all small mammal species he was trapping in upland habitat were found in riparian sites, whereas five species were found in riparian sites only. McGarigal and McComb (1992) found no difference in avian species richness, but species composition varied greatly between riparian and upland habitat.

In the 1980s and 1990s amphibians and reptiles have also been recognized as important components of terrestrial ecosystems. Studies in the U.S. Pacific Northwest have shown that most species of amphibians require riparian or wetland habitat for at least one part of their life cycle (Aubry and Hall 1991), and require cool, moist conditions for maintenance of their respiratory functions. In the western United States, 15% of amphibian species are classified as threatened (Blaustein et al. 1994). There are 18 species of amphibians in British Columbia, 6 of which are threatened by the loss of the riparian and wetland habitats on which they depend (Orchard 1990).

The use of riparian or wetland buffer strips was first proposed as a means of protecting fish habitat from excessive disturbance (Meehan 1991). In 1897 the U.S. government established National Forest Reserves 'for the purpose of securing favorable conditions of water flows' (FEMAT 1993). There are now extensive studies on the effectiveness of buffer strips in reducing the impact of forest practices on aquatic environments, especially streams (Belt et al. 1992). Belt et al. (1992) found many studies on the role of riparian vegetation in controlling stream temperature, stabilization of stream banks, and providing coarse woody debris to maintain channel morphology, stability, and fish habitats. However, few studies have evaluated wildlife use of buffer strips (Darveau et al. 1994). Studies on how to determine buffer strip widths and their effectiveness as filters are also very limited (Belt et al. 1992), and there is very little in the literature on the advantages of varied-width buffers compared with fixed-width buffers. There are, however, a number of papers describing models that can be used to design buffer strips (Belt et al. 1992).

Ecological Principles

Riparian Areas, Wetlands, and Fish

Riparian vegetation plays a key role in the control of water temperature in small and medium-sized streams. This relationship is less important in large rivers and lakes, where riparian vegetation shades less of the wetted surface (Franklin 1991). In smaller streams, incoming radiation may be reduced up to 85% by canopy closure (Wilzbach 1989). Increasing or decreasing the amount of streamside vegetation will affect stream organisms because their basic metabolic processes are temperature-dependent (Miller 1987).

The amount of dissolved oxygen and the waste assimilation capacity of a stream are also influenced by temperature (Miller 1987; McGurk 1989). In low-elevation, low-gradient streams, increased summer temperatures caused by canopy removal may reach levels lethal for fish. Summer temperatures in the U.S. northwest increased from 2°C to 10°C after the streamside canopy was removed (Belt et al. 1992). The Carnation Creek study found that daily stream temperatures increased from 1°C to 5°C following clearcutting (Hartman and Scrivener 1990). Even in mountainous headwater streams, the temperature may increase enough to cause a species shift favouring groups other than salmonids (Wilzbach 1989).

The population viability of some species of trout is directly related to the amount of stream shaded by riparian vegetation. Salmon, brown trout (*Salmo trutta*), and brook trout (*Salvelinus fontinalis*) may die or be outcompeted and replaced by warmer-water species if temperatures exceed 18°C (McGurk 1989). Besides competition from warmwater fish, disease susceptibility increases with elevated temperatures (Brazier and Brown 1973; Gregory and Ashkenas 1990). Also, a shift in species and age composition occurs when shading is decreased (Bisson and Sedell 1982). Brook trout production potential decreased almost 85% when surface shade was decreased to 35% (Hollingsworth 1988), and the abundances of trout, steelhead (*Oncorhynchus mykiss*) smolts, and sculpins decreased after logging. The only stream-rearing species sampled whose numbers remained unchanged after logging was cutthroat trout (*Salmo clarki*) (Hartman and Scrivener 1990).

Not all species react unfavourably to decreased shade, however. Aho and Hall (1976) found that estimated cutthroat trout biomass was greater in the clearcut versus the forested section of a stream. Narver (1972) proposed the 'increased temperature' hypothesis, in which increased winter temperatures caused early emergence of fry, increased appetite, and a longer growing season. Osborne (1980) and Hartman and Scrivener (1990) appear to support this hypothesis. However, even though coho demonstrated earlier fry emergence, a longer growing period, and approximately doubled smolt numbers, the mean size of smolts decreased because two-year-old smolts became rare, spawning adults decreased, and egg-to-fry survival decreased after logging (Hartman and Scrivener 1990).

Canopy closure may also reduce heat loss during winter (Wilzbach 1989). Hewlett and Fortson (1983) found that removal of riparian vegetation could result in about a 5.6°C decrease in winter stream temperatures. The higher winter temperatures that are maintained because of the retention of riparian vegetation may promote some growth in salmonids (Wilzbach 1989).

Riparian vegetation also plays a key role in aquatic primary production by controlling stream temperature and light availability. The amount of streamside vegetation determines whether a stream is autotrophic (dominated by primary production) or heterotrophic (predominantly based on detrital decomposition); this influences the composition of the aquatic invertebrates and therefore the type of prey available to the fish community (Swanson et al. 1982). Melton et al. (1983) described how a stream changed its energy base as it progressed downstream from its headwaters to its mouth. Smaller-headwater streams tend to be heterotrophic (because of the large amount of water surface shaded by vegetation), with the main energy source consisting of organic matter from riparian vegetation. Midsized streams are generally autotrophic (because of their greater surface area exposed to sunlight), with their main energy source being algae and aquatic plants. Large rivers tend to revert to heterotrophy because poor light penetration and turbid waters prevent, to a large extent, photosynthesis from occurring (Melton et al. 1983); consequently, large rivers depend upon their tributaries for input of organic matter (Jahn 1978).

Hollingsworth (1988) stated that 70-80% of the food energy needed by aquatic communities is thought to be supplied by organic matter originating from riparian vegetation (mostly leaf drop) in small to medium-sized streams. Conversely, Swanson et al. (1982) stated that algae and diatoms produced in open-canopy streams make a large contribution to the aquatic energy base and may provide an even more important supply of organic detritus for the macroinvertebrate community than streamside vegetation. However, the Carnation Creek study found that logging reduced the stream's ability to retain leaf litter, and reduced the amount of leaf litter available to 27% of pre-logging levels (Hartman and Scrivener 1990). The study also found that the deciduous leaf litter in the unlogged tributaries was unable to compensate for the loss of leaf litter in the main channel.

Myren and Ellis (1982) found that evapotranspiration by vegetation played a significant role in regulating the amount of summer stream flow even in areas of high annual rainfall. Initially, clearcutting decreased the amount of evapotranspiration and therefore increased the amount of summer flow, but as forest succession proceeded, the authors predicted that increased transpiration would reduce stream flow. The intermediate stage of forest succession was found to have the highest rate of evapotranspiration and transpiration. However, the Carnation Creek study failed to show any increase in annual water yield following canopy removal (P.J. Tschaplinski, pers. comm., 1996). Myren and Ellis (1982) found that reduced summer stream flows resulted in decreased habitat available for fish and invertebrates, and resulted in increased stream temperatures. Studies have shown that decreased stream flow in the summer can effect salmonid productivity (Myren and Ellis 1982). The amount of water actually flowing through a watershed is a result of rainfall and other factors such as terrain, rock type, soils, vegetative cover, and erosion processes within that watershed (Kostadinov and Mitrovic 1994). Each watershed is therefore unique, in that the same amount of rainfall can result in very different stream flow levels. Kostadinov and

Mitrovic (1994) also found that if there was sufficient forest cover in a watershed, it would produce a balanced regime of water flow, preventing large fluctuations in flood or drought conditions.

The root systems of riparian vegetation stabilize lake, river, and stream banks, and thus reduce erosion and sediment input to the water system (Gregory and Ashkenas 1990). Herbaceous plants and shrubs develop dense, stabilizing root systems very quickly. However, tree roots are also needed for long-term stability because they withstand flood conditions more efficiently than shrubs (Wilzbach 1989). Large organic debris (LOD) also plays a role in bank stability and stream channel morphology. In-stream LOD reduces stream velocity and therefore reduces the hydraulic energy that can cause the bank and stream beds to erode (Hamilton 1991). In addition, LOD in the stream channel can form stable pools through logjams, and can control the distribution of sediment and organic matter (Jahn 1978; Wilzbach 1989). House and Boehne (1987) found that stream reaches that had not had their LOD removed had the highest frequency of pools and abundance of off-channel habitat, and the greatest salmonid use. Conversely, stream reaches that had been cleaned of their LOD had the lowest frequency of pools, the least pool area, and the lowest salmonid use. Potts and Anderson (1990) found that pools created by LOD provided over 60% of the total sediment storage in all their study reaches. The Carnation Creek study found that where logging to the stream bank had occurred, LOD in the stream was reduced to approximately 30% of pre-logging amounts (Hartman and Scrivener 1990); small-debris loads became larger, and straight, shallow glides developed along the stream. Murphy and Koski (1989) in Alaska found that 90 years after logging, the LOD volume in a stream remained 70% lower than pre-logging levels. They calculated that 250 years would be needed to recover the original volume in the stream.

Increased streambed erosion, scouring, and bank instability all contribute to increased sediment loads in the stream. Sediment can fill in essential fish habitat such as pools, elevate stream and lake bottoms, and cover important spawning and food-producing substrates (Hollingsworth 1988). Carnation Creek clearly showed that sediment movements and erosion resulted in a straighter, shallower stream with less fish habitat (Hartman and Scrivener 1990). A study in Alaska indicated that there was a great deal of natural variation in sediment levels and that logging could not be directly related to changes in sediment loads (Sheridan et al. 1982). Stream-specific patterns did emerge, however. Sediments in two study sites gradually increased after logging. In one study site, sediment loads upstream decreased, probably due to sediments moving from one site downstream to the other (0.5 km downstream). Shapley and Bishop (1965) found that most fine sediments travelled approximately 0.5 km downstream and then settled to the bottom. Sheridan et al. (1982) stated that research has shown that

Slides contributing to excessive bedload in streams can result from improper drainage from roads or erosion of the toe of a previously stable slope. This erosion may be caused by increased meandering as a result of upstream debris inputs from logging activities. (Photograph by Al Chatterton)

sediment loads generally increase following road construction and logging. Road construction and roads adjacent to riparian areas are considered the main contributors of sediment to a stream (Gregory and Ashkenas 1990; Franklin 1991; Belt et al. 1992).

Increases in sediments, both suspended and on the streambed, can cause problems for many aquatic organisms. The survival of fish eggs and fry, and the feeding success of visual predators such as fish and the invertebrates they feed on, can be severely affected by increases in fine sediments (Miller 1987). Spawning fish may also be responsible for major sediment movements in streams. Fine sediments stirred up during redd development travel downstream, so redds are usually free of such sediments immediately following spawning (Meehan 1991). However, porosity of the redd decreases over time as fine sediments settle on the substrate (Meehan 1991). Low permeability within the stream's substrate caused by excess fine sediments can result in poor oxygen distribution to eggs, and their subsequent suffocation (Reiser and Bjorn 1979). Excess sediment deposition may bury alevins and hinder the emergence of fry (Reiser and Bjorn 1979).

Like changes in temperature, changes in channel morphology associated with logging (such as structural simplification) may influence the species and age compositions of stream fish (Bisson and Sedell 1982). Certain species and age classes such as coho fry (less than one year old) and yearling cutthroat and steelhead prefer pool habitats associated with extensive cover. Bisson and Sedell (1982) found that after logging, these species and age classes decreased, and that, on average, pools in logged streams also decreased in frequency and volume. The study at Carnation Creek indicated that although pre-logging channel morphology changed annually, the change accelerated after logging (Hartman and Scrivener 1990). This included increased streambed erosion and deposition rates.

An important component of the riparian system consists of its connections with the rest of the watershed (Ewel 1978). Because of these connections, however, the riparian area is susceptible to cumulative effects emanating from the entire watershed (Meehan 1991). Such effects (the additive affects of management practices) often demonstrate 'threshold behaviour': the overall effects are minimal up to a certain critical threshold, above which major changes will occur (Franklin 1991).

Cumulative effects can occur over a range of time periods from immediately following forest-harvesting practices to tens or hundreds of years afterward (Meehan 1991). They usually fall into one of the following

Road failures upslope can cause considerable fine sediment input into lakes and streams. (Photograph by Al Chatterton)

categories: (1) changes in the size or timing of rainfall runoff, (2) changes in stream bank stability, (3) changes in the amount of sediment delivered to streams, (4) changes in the distribution of sediment stored in the stream (usually associated with large organic debris), and (5) changes in the energy relationships between water temperature, snowmelt, and freezing (Meehan 1991). The most common cumulative effects are associated with road construction and hydrological processes (Franklin 1991). The cumulative effects of rain on snow events may cause severe flooding. During these events, roads and clearcutting can increase peak stream flows compared with flows in areas of unroaded and unlogged forests (Franklin 1991). In the Queen Charlotte Islands of British Columbia, logging, vegetation removal, and road construction have been shown to increase the frequency of landslides: logging by a factor of 34 times, vegetation removal by a factor of 31 times, and road construction by a factor of 87 times (Bunnell et al. 1995).

Riparian Areas, Wetlands, and Invertebrates

Riparian vegetation plays a large role in determining the composition of the aquatic invertebrate community associated with a particular riparian or wetland environment. Aquatic insect communities may consist of various combinations of collectors, shredders, grazers, predators, and parasites (Melton et al. 1983). Collectors feed on small particulate matter broken down from vegetation (such as leaf litter) and algae detritus; they are the principal prey for freshwater salmonids (Melton et al. 1983). Shredders feed on leaves, twigs, and other plant tissue. Grazers, also called scrapers, feed on filamentous algae and attached aquatic plants. Grazers are also important prey species for freshwater salmonids. Predators and parasites feed on other living organisms (Melton et al. 1983).

In small headwater streams with a dense overhanging canopy, the dominant invertebrates are collectors and shredders, with a small component of grazers and predators (Melton et al. 1983). In midsized streams (which tend to have less canopy closure), the dominant invertebrates are collectors and grazers, with a small component of shredders and predators. In large rivers (with very little canopy closure and turbid waters), the dominant invertebrates are collectors, with a small component of predators. Invertebrate parasites may be found anywhere in a riparian area or wetland (Melton et al. 1983).

Hachmoller et al. (1991) found that habitat alterations caused shifts in the species composition of invertebrate communities: when headwaters were logged and channelized so as to be similar in morphology and habitat to midsized streams, the aquatic invertebrates also shifted to those found in that habitat. Likewise, if a riparian area was logged and channelized so that stream velocity decreased to that found in lower reaches, the invertebrate population shifted to those found in that habitat.

The type of litter affects the quantity of food available to shredders and collectors. Quickly decomposing litter (e.g., alder, salmonberry) breaks down into fine particulates and supplies food in the fall and winter for shredders and collectors. Slowly decomposing litter (e.g., grasses, sedges, conifers, conifer litter) breaks down and provides food in the spring and summer for shredders and collectors (Wilzbach 1989).

Several studies have shown that clearcuts can lead to short-term increases in invertebrate densities in streams. However, although such densities increased in most sites, Erman and Mahoney (1983) found that they decreased significantly in logged streams compared with unlogged ones, or in streams with buffer strips. Even streams logged 10 years before sampling showed a significant decrease in diversity compared with unlogged controls. Three study sites showed decreases in both density and diversity. The Carnation Creek study also demonstrated decreases in densities of aquatic invertebrates after logging of the stream bank compared with the same area before logging (Hartman and Scrivener 1990). Aquatic invertebrate densities also decreased in areas with a buffer strip, compared with pre-logging levels.

Newly emerged damsel flies cling to riparian vegetation until their wings strengthen before flight. Both the aquatic and aerial stages of this insect provide food for fish and numerous riparian species. (Photograph by Steve Voller)

Most aquatic insects spend their terrestrial or aerial adult stage in the riparian zone (Jackson and Resh 1989); dispersal, feeding, and mating take place on land. The terrestrial environment is therefore critical for survival. In a study of adult aquatic insects, Jackson and Resh (1989) found that mating usually took place at a defined time and place within the riparian area. Abundance was greatest near the stream and decreased with increasing distance from the stream. However, some species were still common 150 m from the stream.

Both aquatic and terrestrial invertebrates are important prey for stream-dwelling salmonids. A significant proportion of the terrestrial invertebrates available to these fish consists of the adult, aerial stages of aquatic species. Hollingsworth (1988) noted that terrestrial insects may account for more than 40% of a trout's summer diet in a small stream. Tschaplinski (1987) stated that more than two-thirds of the insect diet of stream salmonids may be terrestrial. Predators such as coho feed largely on aquatic and terrestrial invertebrates that drift downstream with the current. In Carnation Creek, drift samples collected at 3-hour intervals over 24-hour periods revealed that 75-91% of all stream drift and 60-84% of all estuarine drift consisted of aquatic invertebrates (Tschaplinski 1987). Correspondingly, 59-78% of the diet of coho fry and yearlings of stream coho, and 57-77% of the diet of estuarine coho, consisted of aquatic invertebrate drift (Tschaplinski 1987). The balance of both the drift and the coho diet in the freshwater and estuarine reaches of Carnation Creek consisted of terrestrial invertebrates. Besides supporting the fish population, both aquatic and terrestrial insects support insectivores (such as birds and bats) found in the riparian zone (Jackson and Resh 1989).

Riparian areas and freshwater wetlands support other invertebrates besides insects. A major group of freshwater invertebrates are the molluscs, found in rivers and wetlands; they are mainly benthic and are often associated with aquatic vegetation (Clark 1978). Crustaceans are another major group; they can be found in all portions of a freshwater wetland and in many rivers and streams. Crayfish are a common crustacean in rivers and wetlands (Clark 1978).

Riparian Areas, Wetlands, and Herpetofauna

Many studies have indicated a decline in amphibian populations and an increase in reptile populations in managed stands (Corn and Bury 1991). Reptiles such as lizards often increase in numbers and take advantage of new forest openings (Corn and Bury 1991). While doing a comparative study between reptiles and amphibians, Welsh and Lind (1991) found that species richness did not vary between forest age classes, but composition did. They found that old, wet sites contained relatively more amphibian species, and young, dry sites contained relatively more reptilian species. Snakes seldom require aquatic habitat to fulfil their life cycles, but are often found in aquatic environments, especially garter snakes (Melton et al. 1983).

In British Columbia, six species of amphibians (tiger salamander [*Ambystoma tigrinum*], Pacific giant salamander [*Dicamtodon tenebrosus*], Coeur d'Alene salamander [*Plethodon idahoensis*], tailed frog [*Ascaphus truei*], Great Basin spadefoot toad [*Scaphiopus intermontanus*], and the northern leopard frog [*Rana pipiens*]) and two species of reptiles (painted turtle [*Chrysemys picta*] and night snake [*Hypsiglena torquata*]) are on the Red or Blue List; they depend on riparian areas for part or all of their life cycle (Stevens et al. 1995).

Habitat requirements are often complex, and specific conditions at specific times may be required to ensure breeding success. For example, Richter (1993, in FEMAT 1993) found that northwestern salamanders (*Ambystoma gracile*) attach their eggs to aquatic vegetation at a specific height below the water surface; should this water level change prior to hatching, partial or complete reproductive failure may result. Studies have shown that amphibian species such as tailed frogs breed in cascading headwater streams (Corn and Bury 1991) and have a narrow temperature tolerance (Bury et al. 1991). Pacific giant salamanders may also breed in cascading headwater streams and in ponds and backwaters of streams, along with species such as the red-legged frog (*Rana aurora*), northwestern salamander, and roughskin newt (*Taricha granulosa*). Pacific giant salamanders may, however, have wider temperature tolerances than any of the other species (Bury et al. 1991).

Many species may also breed in ponds that are too small to be included in wetland management plans for conservation (FEMAT 1993). Because of the poor dispersal capabilities that most amphibian species are assumed to have, the upland habitat these species use may depend on their proximity to water (Corn and Bury 1991). Blaustein et al. (1994) suggested that, because amphibians are assumed to have poor dispersal capabilities, relative lack of mobility, and strong site fidelity, they would have a poor chance of recolonizing areas of local extinction if their habitat became fragmented.

Aubry and Hall (1991) in southern Washington found that the adult, terrestrial stage of tailed frogs and red-legged frogs were most abundant in mature stands rather than young stands. The adult, terrestrial stage of northwestern salamanders and roughskin newts were most common in old growth. The juveniles of these species disperse every year from ponds and creeks to their preferred habitats (Aubry and Hall 1991). Corn and Bury (1991) in Oregon found that northwestern salamanders and tailed frogs were captured in old growth but not clearcuts. Conversely, all species captured in clearcuts were also captured in old growth. However, the total abundance was not significantly different between the two habitat types.

Corn and Bury (1991) pointed out that there are few studies on the long-term effects of logging on amphibians. However, studies by Bennett et al.

Red-legged frogs seek cover in aquatic vegetation on the edges of streams, ponds, lakes, and wetlands. (Photograph by Steve Voller)

(1980, in Corn and Bury 1991) and Pough et al. (1987, in Corn and Bury 1991) demonstrated a reduction in terrestrial amphibians in eastern U.S. second-growth forests. Studies by Bury (1983) in the redwood forests of California indicated that logging had a long-term effect on resident amphibians, causing reduced abundance and composition change. In addition, sites sampled 6 to 14 years after logging had still not recovered to their original levels of amphibian abundance and composition.

Corn and Bury (1991) in Oregon noted a decline in density and biomass of stream amphibians in logged stands 14 to 40 years old. In a study on Vancouver Island, Dupuis (1993) found that clearcutting could reduce amphibian populations by up to 70%. The study also indicated that salamander densities in second-growth stands were highest within 10 m of a stream, and their densities in these areas were roughly equal to those found randomly throughout old growth.

Orchard (1990) summarized the adverse effects of sedimentation from logging on the mountain streams that support tailed frogs and Pacific giant salamanders. When silt covers the stream rocks, the tadpoles of tailed frogs are unable to hold on to the rock surfaces and are washed away in the current. Pacific giant salamanders also avoid warm, cloudy waters, such as those with a heavy sediment load (Orchard 1990). In British Columbia, Licht (1971) and MacIntyre and Palermo (1980) proposed that the greatest threats to red-legged frogs were habitat destruction and competition from introduced species.

Canada geese (and other waterfowl) use wetland and riparian areas for protection, nesting, and feeding. (Photograph by Steve Voller)

Riparian Areas, Wetlands, and Birds

In British Columbia, 48 bird species on the Red and Blue lists use riparian or wetland areas for part or all of their life cycles (Stevens et al. 1995). Examples include the American White Pelican (*Pelecanus erythrorhynchos*), MacFarlanes Western Screech Owl (*Otus kennicottii macfarlanei*), Sharp-Tailed Sparrow (*Ammospiza caudacuta*), Western Grebe (*Aechmophorus occidentalis*), American Bittern (*Botaurus lentiginosus*), Eared Grebe (*Podiceps nigricollis*), Long-eared Owl (*Asio otus*), and Mourning Warbler (*Oporornis philadelphia*) – all of which use riparian areas or wetlands for part of their life cycles and are thought to be endangered due to loss of this habitat (B.C. Ministry of Environment, Lands and Parks 1992). As well, American Dippers (*Cinclus mexicanus*) (the only truly aquatic songbird) and Northern Waterthrushs (*Seiurus noveboracensis*) (one of the few forest bird species found only in riparian areas) are totally dependent on riparian areas for their survival (Terres 1991; J.P. Savard, pers. comm., 1996).

The diversity of bird species found in a riparian zone is influenced by the multiple edge effects caused by the gradient of changing vegetation from the water's edge to upland habitat (Bull 1978). Food for birds in a riparian area or wetland includes aquatic vegetation, invertebrates (insects, molluscs, crustaceans, and so on), reptiles, amphibians, mammals, and other birds. Waterfowl use these habitats for protection, nesting, and feeding (Melton et al. 1983). Ryder (1951) found that waterfowl production was directly

Wildfire and other small-scale disturbances create canopy gaps that are a primary source of heterogeneity in forest structure and composition. Canopy gaps enable species to become established that could not do so under a closed canopy. Large wildfires and other large-scale disturbances, however, can be of great magnitude over large areas, often returning stands to early seral stages. (Photograph by John Parminter)

Fragmentation poses a serious threat to biological diversity today. Fragmentation occurs in three phases: (1) the landscape is perforated by the cutting units, (2) the cutover areas begin to coalesce and the remaining forest is fragmented into isolated patches, and (3) the second growth now forms the matrix with only small islands of uncut forest remaining. (Photograph by Joan Voller)

Species perceive connectivity at different spatial and temporal scales. A cougar (*Felis concolor*) (right) may have a home area of 40,000 ha and move 100 km in a day, whereas a shrew (*Sorex* spp.) (below) may live its life within 10 ha and move less than 1 km per day. (Photograph by Scott Harrison)

Increased amounts of edge enhance populations of edge species such as the Columbian black-tailed deer. However, increased amounts of edge also decrease populations of other species. Large amounts of edge are also associated with a highly fragmented landscape. (Photograph by Joan Voller)

Riparian processes at work: stream banks stabilized by root networks of mature trees and large organic debris; leaf input providing food for grazing invertebrates as well as nutrients into the water; riparian vegetation (rose hips) providing food for birds and small mammals utilizing riparian habitats; and large streamside trees providing nesting habitat for small birds and perch trees for birds of prey hunting edge species. (Photograph by Joan Voller)

Interior habitat is generally defined as that portion of habitat that does not experience edge effects and thus maintains its functional viability for plant and animal communities. The amount of interior habitat is determined by the size and shape of that habitat and by the edge variable being measured. (Photograph by Scott Harrison)

Much of British Columbia's biodiversity is associated with old-growth forests. Recognizing which species are associated with old growth and the nature of that association is important in managing for biodiversity. (Photograph courtesy B.C. Ministry of Forests)

Large woody debris may remain in one place throughout its decay cycle or may move throughout a system providing various functions before it ultimately decays or exits the system. (Photograph by Steve Voller)

proportional to the amount of edge between the terrestrial and aquatic environments. Nummi et al. (1994) and Merendino and Ankney (1994) found that Mallard (*Anas platyrhynchos*) habitat preference was associated with invertebrate density and the amount of cover in a wetland. Savard et al. (1994) found that pond size and productivity was the most important factor influencing waterfowl densities. The amount of marsh and riparian vegetation also affected species composition (Savard et al. 1994).

Many small birds use the brushy riparian vegetation as cover from predators. Gill et al. (1974) noted that riparian areas in the central Appalachians exhibited a higher bird density than non-riparian habitat. Hooper (1967) noted 44% more breeding birds and 33% more bird species in riparian areas than midslope sites in the American southeast. Migrating passerines in the American southwest have demonstrated a preference for riparian rather than non-riparian habitat (Stevens et al. 1977). Stevens et al. (1977) stated that, in Arizona, more than twice as many breeding birds (individuals and species) were found in riparian habitat. They also noted that poor adjacent habitat forced more birds into the riparian habitat.

McGarigal and McComb (1992) noted that, in Oregon, species richness did not differ between riparian areas and unmanaged midslope habitat, but species composition did. They cautioned that most studies identifying riparian areas as preferred were undertaken in relatively arid environments. They found that when comparing mid-elevation riparian areas with mature, unmanaged upslope areas, bird species abundance, richness, and diversity were greater in the upslope sites. The authors also recognized the limitations of their study, which was limited to mature, unmanaged forests with a greater number of snags and large conifers in the upslope areas; other studies found that streamside versus upslope affiliations may vary along the riparian elevational gradient (e.g., Finch 1991a). Snags, a common component of natural riparian areas, are essential habitat for cavity-nesting birds (Bull 1978).

Best and Stauffer (1980) found that nesting success in riparian zones, ranging from open fields to closed-canopy woodlands, was hindered most commonly by predation. Causes of nest failure, in order of significance, were predation by birds, snakes, and small and large mammals; desertion; Brown-headed Cowbird (*Molothrus ater*) parasitism; and natural disasters. For the smaller bird species, most nest failures were due to large-mammal predation and cowbird parasitism; the higher the nest, the greater the chance of nesting success (Best and Stauffer 1980). Nests in the open were more vulnerable to predators and poor environmental conditions than those in more sheltered sites (Best and Stauffer 1980). Because birds are vulnerable while nesting, the nesting habits of species may influence their success or failure during human disturbance; species that are solitary nesters, such as Bitterns, have a much lower tolerance to disturbance than colony nesters (Landin 1979).

Riparian Areas, Wetlands, and Mammals

In British Columbia, 34 mammalian species on the Red and Blue lists use riparian areas for all or part of their life cycles (Stevens et al. 1995). Species such as river otters (*Lutra canadensis*), beavers (*Castor canadensis*), and muskrats (*Ondatra zibethica*) are riparian or wetland-dependent. Bats are strongly associated with aquatic habitats because aquatic insects are a major component of their diet (FEMAT 1993). In B.C., Red- and Blue-listed species such as the southern red bat (*Lasiurus borealis*) and the Pacific water shrew (*Sorex bendirii*) are thought to be endangered because of loss of riparian or wetland habitat, on which they depend for part of their life cycle (B.C. Ministry of Environment, Lands and Parks 1992).

As mentioned in the previous section, most studies comparing riparian and upland use were undertaken in relatively arid regions. However, Cross (1985) looked at more mesic sites in comparing small-mammal use of riparian, transition, and upland sites. Such use was significantly higher in riparian areas: all species captured in the three sites were present in the riparian site, but not all species were found in the other two sites.

In the Cascade Range, Doyle (1990) also found that small-mammal abundance and richness was greater in the riparian site than in the upland site. Most species weighed more, and more adults were in breeding condition in the riparian areas. These results indicated that the montane riparian areas provided superior habitat for many small mammals. This superiority may have been due to such factors as greater availability of water and forage (Doyle 1990). Small mammals forage on conifer seeds, mycorrhizal fungi, and insects found in the riparian zone (Simons 1985). In the central Coast Ranges of Oregon, however, McComb et al. (1993) found that the number of small mammals captured did not differ between streamside and upslope sites.

Many mammals are attracted to riparian areas because they provide thermal cover, foraging habitat, and cover for hiding, resting, breeding, and rearing. The riparian vegetation provides amenable microclimates by buffering temperature and humidity extremes (Gregory and Ashkenas 1990). Doyle's (1990) study indicated that temperatures were lower and more stable in riparian versus upland sites, which could help reduce the energy expended for thermoregulation. In addition, the soils found in riparian areas tend to be more friable and coarse than in upland sites, making it easier for small mammals to burrow (Doyle 1990). Compton et al. (1988) found a significant relationship between white-tailed deer (*Odocoileus virginianus*) distribution and the amount of riparian vegetation in Montana; therefore the amount of riparian cover could determine the number of deer a bottomland habitat could support.

Design and Function of Riparian and Wetland Buffers

There are several recognized and documented functions of riparian and wetland buffer strips. They

- provide limited sinks for sediments, nutrients, and other chemicals from upslope
- provide limited control of surface and subsurface water flow
- maintain stable banks and limit erosion (Belt et al. 1992) and provide shade to control temperature in streams (Nutter and Gaskin 1988)
- provide a food source for aquatic dwellers in the form of vegetation debris and terrestrial invertebrates (Erman and Mahoney 1983)
- provide a source of coarse woody debris (Hamilton 1991)
- provide shelter, forage, and breeding habitat for a diverse assemblage of animal species
- provide habitat for a diverse assemblage of plant species.

Buffer strips may be established to trap sediments. Belt et al. (1992) stated that the main factors controlling the movement of sediments (via overland flow) within a buffer strip were slope and the density of obstructions such as vegetation, rocks, and coarse woody debris. Clinnick (1985) stated that the vegetation in a buffer strip spreads the water flow and reduces velocity, but if the buffer is not wide enough, the area becomes saturated and sediment is carried to the water. Studies have indicated that where the slope is steeper, buffer strips should be wider (Belt et al. 1992). Van Groenewould (1977) suggested that in order to trap sediments, buffers should range from 15 m to 65 m, depending on the steepness of the terrain. Studies in Idaho have also shown that sediments can move overland up to 90 m within a buffer strip, but the buffer strips are not effective in controlling channelized flow originating outside the buffer: if obstructions are removed from within the buffer, the overland flow may travel further than 90 m (Belt et al. 1992). Toews and Moore (1982) found that where logging occurred to the stream bank, stream bank erosion increased 250% from the pre-harvest condition. However, if a buffer strip of approximately 5 m was left, streambank erosion increased only 32%.

The role of buffer strips as moderators of stream temperature is well documented (Belt et al. 1992). The removal of streamside vegetation can lead to increases in stream temperature, as noted under 'Riparian Areas, Wetlands, and Fish' earlier in this chapter. However, McGurk (1989) has shown that the width of a buffer strip is not a good determinant of stream temperature; the density of canopy actually shading the stream is a better determinant. Brazier and Brown (1973) stated that the width of a buffer alone does not determine stream temperature; however, width is related to effectiveness through a complex relationship between canopy density, height, stream width, and stream discharge. They found that the density of the canopy along the incoming path of radiation was most closely related to stream temperature. In addition, Swift and Baker (1973) found that the discharge of groundwater had more influence on cooling a heated stream once it entered the buffer than the shading from the canopy did. Brown et al. (1971) also stated that shade alone could not be counted on to cool a stream, as most of the cooling was done by the influx of groundwater in shaded areas.

Erman et al. (1977), Erman and Mahoney (1983), and Newbold et al. (1980) noted the effects on stream macroinvertebrates in riparian areas with and without buffer strips. All three studies found that aquatic invertebrate communities shifted and diversity decreased when logging occurred to a stream edge. Erman et al. (1977) also noted that buffer strips of 30 m or less were insufficient to protect benthic invertebrates from logging impacts, whereas 30 m or more were sufficient.

McDade et al. (1990) found that a 30 m buffer could supply to the stream 85% of the large organic debris that a natural stand could, but a 10 m buffer could provide only 50% or less.

In a study of birds occupying a lakeshore buffer, Johnson and Brown (1990) surmised that the species they studied would be able to breed if the buffer were 75 m wide, with the first 25 m untouched and the rest with more than 50% canopy cover remaining. However, they qualified their remarks by stating that the actual width needed for breeding birds was unknown. Triquet et al. (1990) found that bird species richness and diversity were greater in an uncut, mature forest and a harvest unit with a riparian buffer strip than

Buffer strips along stream courses can provide habitat required by many species dependent on the riparian transition area. (Courtesy B.C. Ministry of Forests)

in a clearcut without a buffer strip. They found that many species occurred in the riparian buffer, including some interior forest species. However, they also found that some interior forest birds would not use buffers, and that these species decreased in abundance in the mature forest adjacent to a clearcut. Moreover, because of their linear nature buffer strips are greatly influenced by the edge effect between the forest and clearcut (Yahner 1988). (For more information on edge effects, refer to Chapter 8.) Studies have also indicated that buffer strips less than 100 m wide are insufficient for maintaining many terrestrial wildlife populations (Darveau et al. 1993).

Belt et al. (1992) noted that there were many studies on the cost-effectiveness of buffer widths, but warned that these must be speculative as they were based on cost-effectiveness criteria involving values not reflected in the market, such as the value of buffer strips' contribution to preserving biodiversity. Dykstra and Froehlich (1976) studied the economics of three alternatives that would leave streams clean following timber harvest. No one method proved to be more economical overall. In four of their study sites, the most cost-effective was a 55-foot buffer; in three sites, it was conventional felling followed by stream clean-up; and in another three sites, it was cable-assisted logging. They were quick to point out, however, that not all methods protect the stream or the terrestrial component of the riparian area to the same degree, and that methods must not be chosen based on cost alone.

Belt et al. (1992) found no studies that discussed the value of variable-width buffer strips over fixed-width buffer strips. They pointed out that variable-width buffers allow the riparian zone to better mimic a natural situation, although they are more complex to implement. However, Gregory and Ashkenas (1990) felt that variable-width buffer strips took into consideration topographic irregularities and thus simplified harvest unit layout.

Washington, Oregon, and California all prescribe the use of variable-width buffers (which follow ecological boundaries instead of being set at arbitrary distances from lakes, rivers, or streams), and design them using stream classification and site-specific factors (Belt et al. 1992). Franklin (1991) pointed out the need to recognize the longitudinal nature of streams and rivers, taking into consideration sources of sediment, large organic debris, and other factors associated with the aquatic system.

Research Needs

In order to understand the ecosystem as a whole, we must understand all the parts of the riparian system: 'understanding the distribution of biota among different systems and assessing their functional roles are key both to understanding how ecosystems function and to their conservation' (Ray 1992). The following are the major information gaps on riparian areas and wetlands identified in the literature.

Classification systems. A comprehensive classification system for riparian ecosystems and wetlands in British Columbia needs to be completed (A. Banner, pers. comm., 1996).

Coarse woody debris. More research is needed to determine the amounts of coarse woody debris (large organic debris) needed in streams, and how much mature timber is needed in order to provide a sustainable amount (Hamilton 1991).

Role of adult aquatic insects in the riparian zone. We need to understand the factors that may effect the function, distribution, and abundance of adult aquatic insects in the riparian zone as this stage is vital for the continued existence of most aquatic insects (Jackson and Resh 1989).

Baseline knowledge about invertebrate species. At present, invertebrates cannot be nominated for endangered or threatened status (Cannings 1990). This is in the process of being changed, and species known to be rare may soon be given proper status under the B.C. Wildlife Act and the Forest Practices Code (S. Cannings, pers. comm., 1996). However, because we lack baseline knowledge about the populations and distributions of the various species, many important species may be missed. Cannings (1990) feels that we have yet to identify over half of the invertebrate species found in B.C.

Effects of fragmentation on amphibians. Studies to determine these effects should be initiated (Corn and Bury 1991; Blaustein et al. 1994).

Effects of timber harvest on amphibians. Additional research on the abundance and presence of amphibians after timber harvest is needed (Bury et al. 1991). In addition, the buffer-strip widths and forest-patch sizes needed to protect amphibians have to be determined (Bury et al. 1991).

Amphibian census. Strict sampling studies and appraisals of the direction, magnitude, and causes of change in amphibian populations need to be undertaken to determine the seriousness of the apparent decline in these populations (Blaustein et al. 1994).

Natural history information for amphibians. A more thorough knowledge base for amphibian species is needed, including information on life histories, ecology (such as dispersal capabilities), and population sizes (Bury et al. 1991; Corn and Bury 1991). Information on habitat preferences and tolerances to environmental change also needs to be collected (Bury et al. 1991).

Comparisons of fauna in managed stands. Bury and Corn (1988) feel that there is a lack of information on herpetofauna and other wildlife in managed second-growth stands compared with other management regimes (such as partial cutting) and naturally regenerated habitats. They point out that whereas most of the Pacific Northwest will soon be dominated by managed stands, our understanding of species and their ability to cope with these changes is inadequate.

Use of riparian areas outside the breeding season. Very little research has been done on the use of riparian areas outside the breeding season, such as their use during migration, winter roosting, and so on (J.P. Savard, pers. comm., 1996).

Nesting success of birds in the riparian zone. Best and Stauffer (1980) felt that it was important to study the nesting success of various types of birds in different riparian habitats and geographic areas to determine the causes of nest failure.

Neotropical migratory birds. Although many questions about neotropical migrants involve more than the riparian zone, the apparent decline of these birds is a research priority in all habitats that they apparently depend upon (Finch 1991b). Finch (1991b) lists many unanswered questions about neotropical migrants that may be appropriate for study in the riparian zone, under such titles as: analyses of monitoring data and techniques, questions about migrant population trends, effects of land-use practices, the value of migration corridors and stopover sites, the status of migrants in western North America, and demographic and biotic factors limiting populations.

Species inventories. In order to preserve natural diversity in a particular riparian zone, it is necessary to identify the riparian-dependent species that are endemic to the area and that have evolved in that 'geographic domain' (Knopf and Samson 1994). Howard and Allen (1988) suggest that the critical first step in managing a riparian zone is undertaking a fish and wildlife inventory, with emphasis on rare and endangered species.

Effects of logging within a buffer strip. There is a need to understand the effects of partial cutting within the buffer strip on species that use the area (Nutter and Gaskin 1988).

Buffer widths. Hamilton (1991) stated that the actual width of buffer needed for the various stream sizes and types is still unknown. It may be necessary to have wider buffers on laterally active streams in order to provide a sustainable source of coarse woody debris. Nutter and Gaskin (1988) pointed out that there was no scientific way to define the optimal buffer width. However, Knopf and Sampson (1994) pointed out the importance of knowing the widths necessary to buffer riparian bird species from upland disturbances.

Economic costs and benefits of buffer strips. Studies to determine the costs and benefits of buffer strips for fisheries as well as other biodiversity concerns should be undertaken (Hamilton 1991). More research into calculating a dollar value for the less quantifiable factors in the riparian zone – such as wildlife habitat, erosion control, and stream temperature regulation – would aid in determining these costs and benefits (Caulfield et al. 1992).

Monitoring the riparian zone. Monitoring of past and present management practices in the riparian zone should be undertaken to study the effectiveness and validity of practices and enable the development of future policies (Gregory and Ashkenas 1990).

Fragmentation and isolation of the riparian zone. The impact of fragmentation and isolation (discontinuity) on the ecological value of the riparian zone should be studied (J.P. Savard, pers. comm., 1996).

Literature Cited

Agriculture Canada Expert Committee and Soil Survey. 1987. The Canadian system of soil classification. 2nd ed. Agriculture Canada Publication 1646. Ottawa, Ont.

Aho, R.S., and J.D. Hall. 1976. Cutthroat trout populations in clearcut and forested sections of a stream. Bulletin of the Ecological Society of America 57(1):41-42.

Aubry, K.B., and P.A. Hall. 1991. Terrestrial amphibian communities in the southern Washington Cascade Range. Pages 327-339 in L.F. Ruggiero, K.B. Aubry, A.B. Carey, and M.H. Huff (eds.). Wildlife and vegetation of unmanaged Douglas-fir forests. USDA Forest Service General Technical Report PNW-GTR-285. Portland, Ore.

B.C. Ministry of Environment, Lands and Parks. 1992. Vertebrates at risk in British Columbia: the 1992 Red and Blue Lists. Draft. Wildlife Branch, Victoria, B.C. 87 pp.

B.C. Ministry of Forests. 1994. British Columbia Forest Practices Code: Standards. 216 pp.

Belt, G.H., J. O'Laughlin, and T. Merrill. 1992. Design of forest riparian buffer strips for the protection of water quality: analysis of scientific literature. Idaho Forest, Wildlife and Range Policy Analysis Group, University of Idaho, Report No. 8. 35 pp.

Benke, A.C. 1990. A perspective on America's vanishing streams. Journal of the North American Benthological Society 9:77-88.

Best, L.B., and D.F. Stauffer. 1980. Factors affecting nesting success in riparian bird communities. Condor 82:149-158.

Bisson, P.A., and J.R. Sedell. 1982. Salmonid populations in streams in clearcut vs. old-growth forests of Western Washington. Pages 121-129 in Fish and wildlife relationships in old-growth forests: proceedings of a symposium. April 1982, Juneau, Alaska.

Blaustein, A.R., D.B. Wake, and W.P. Sousa. 1994. Amphibian declines: judging stability, persistence, and susceptibility of populations to local and global extinctions. Conservation Biology 8(1):60-71.

Brazier, J.R., and G.W. Brown. 1973. Buffer strips for stream temperature control. Forest Research Laboratory, Oregon State University, Corvallis, Ore., Research Paper 15. 9 pp.

Brown, G.W., G.W. Swank, and J. Rothacher. 1971. Water temperature in the Steamboat drainage. USDA Forest Service RP PNW-119. 17 pp.

Bull, E.L. 1978. Specialized habitat requirements of birds: snag management, old growth, and riparian habitat. Pages 74-82 in R.M. DeGraaf (tech. coord.). Nongame bird habitat management in coniferous forests of the western United States: proceedings of a workshop. 7-9 February 1977. USDA Forest Service, General Technical Report PNW-GTR-64. Portland, Ore.

Bunnell, P., S. Rautio, C. Fletcher, and A. Van Woudenberg. 1995. Problem analysis of integrated resource management of riparian areas in British Columbia. B.C. Ministry of Forests and B.C. Ministry of Environment, Lands and Parks, Victoria, BC. Working Paper 11/1995.

Bury, R.B. 1983. Differences in amphibian populations in logged and old growth redwood forests. Northwest Science 57:167-178.

Bury, R.B., and P.S. Corn. 1988. Douglas-fir forests in the Oregon and Washington Cascades: relation of the herpetofauna to stand age and moisture. Pages 11-22 in R.C. Szaro, K.E. Severson, and D.R. Patton (eds.). Management of amphibians, reptiles, and small mammals in North America: proceedings of the symposium. 19-21 July, Flagstaff, Ariz. USDA Forest Service General Technical Report RM-166.

Bury, R.B., P.S. Corn, K.B. Aubry, F.F. Gilbert, and L.L.C. Jones. 1991. Aquatic amphibian communities in Oregon and Washington. Pages 353-362 in L.F. Ruggiero, K.B. Aubry, A.B. Carey, and M.H. Huff (eds.). Wildlife and vegetation of unmanaged Douglas-fir forests. USDA Forest Service General Technical Report PNW-GTR-285. Portland, Ore.

Cannings, S. 1990. Endangered invertebrates in B.C.. Bioline 9(2):15-17.

Caulfield, J., J. Welker, and R. Meldahl. 1992. Opportunity costs of streamside management zones on an industrial forest property. Pages 639-644 in Seventh Biennial Southern Silvicultural Research Conference. Mobile, Ala.

Clinnick, P.F. 1985. Buffer strip management in forest operations: a review. Australian Forestry 48(1):34-35.

Compton, B.B., R.J. Mackie, and G.L. Dusek. 1988. Factors influencing distribution of white-tailed deer in riparian habitats. Journal of Wildlife Management 52(3):544-548.

Corbett, E.S., J.A. Lynch, and W.E. Sopper. 1978. Timber harvesting practices and water quality in the eastern United States. Journal of Forestry 76(8):484-488.

Corn, P.S., and B.R. Bury. 1991. Terrestrial amphibian communities in the Oregon Coast Range. Pages 305-317 in L.F. Ruggiero, K.B. Aubry, A.B. Carey, and M.H. Huff (eds.). Wildlife and vegetation of unmanaged Douglas-fir forests. USDA Forest Service General Technical Report PNW-GTR-285. Portland, Ore.

Cross, S.P. 1985. Responses of small mammals to forest riparian perturbations. Pages 269-275 in North American Riparian Conference. University of Arizona, Tucson, Ariz.

Dahl, T.E. 1990. Wetland losses in the United States 1780's to 1980's. U.S. Department of Interior, Fish and Wildlife Service, Washington, D.C.

Darveau, M., J. Huot, L. Belanger, and J.-C. Ruel. 1994. Mid-term effects of windfall on bird use of riparian forest strips. Pages 104-109 in I.D. Thompson (ed.). Forests and wildlife ... towards the 21st century, XXI IUGB Proceedings, Halifax, 1993, Vol. 2. Canadian Forest Service. Petawawa, National Forestry Institute, Chaulk River, Ont.

Doyle, A.T. 1990. Use of riparian and upland habitats by small mammals. Journal of Mammalogy 7(1):14-23.

Dupuis, L.A. 1993. The status and distribution of terrestrial amphibians in old-growth forests and managed stands. M.Sc. thesis, University of British Columbia, Vancouver, B.C. 65 pp.

Dykstra, D.P., and H.A. Froehlich. 1976. Costs of stream protection during timber harvest. Journal of Forestry 74(10):684-687.

Erman, D.C., and D. Mahoney. 1983. Recovery after logging in streams with and without bufferstrips in Northern California. California Water Resources Center, University of California, Davis, Calif. 50 pp.

Erman, D.C., J.D. Newbold, and K.B. Roby. 1977. Evaluation of streamside bufferstrips for protecting aquatic organisms. California Water Resources Center, University of California, Davis, Calif. 38 pp.

Ewel, K.C. 1978. Riparian ecosystems: conservation of their unique characteristics. Pages 56-62 in Strategies for protection and management of floodplain wetlands and other riparian ecosystems: proceedings of the symposium. 11-13 December 1978, Callaway Gardens, Georgia. USDA Forest Service General Technical Report WO-12.

Finch, D. 1991a. Positive associations among riparian bird species correspond to elevational changes in plant communities. Canadian Journal of Zoology 69:951-963.

–. 1991b. Population ecology, habitat requirements, and conservation of neotropical birds. USDA Forest Service General Technical Report RM-205. Fort Collins, Colo. 26 pp.

Forest Ecosystem Management Assessment Team (FEMAT). 1993. Forest ecosystem management: an ecological, economic, and social assessment. USDA Forest Service (and other agencies), Portland, Ore.

Franklin, J.F. 1991. Scientific basis for new perspectives in forests and streams. Pages 3.1-3.38 in R.J. Naiman (ed.). New perspectives for watershed management. Springer-Verlag, New York.

Gill, J.D., R.M. DeGraaf, and J.W. Thomas. 1974. Forest habitat management for non-game birds in central Appalachia. USDA Forest Service RN NE-192. Radner, Penn. 6 pp.

Gregory, S. 1976. Primary production in a stream in a clearcut and old-growth forest in the Oregon Cascades. Bulletin of the Ecological Society of America 57(1):41.

Gregory, S., and L. Ashkenas. 1990. Riparian management guide. Willamette National Forest, Ore. 120 pp.

Hachmoller, B., R.A. Matthews, and D.F. Brakke. 1991. Effects of riparian community structure, sediment size, and water quality on the macroinvertebrate communities in a small, suburban stream. Northwest Science 65(3):125-132.

Hamilton, S.R. 1991. Streamside management zones and fisheries protection. Pages 45-49 in Proceedings WildFor91. Wildlife and forestry: towards a working partnership. Canadian Society of Environmental Biologists and Canadian Pulp and Paper Association.

Hartman, G.F., and J.C. Scrivener. 1990. Impacts of forestry practices on a coastal stream ecosystem, Carnation Creek, British Columbia. Canadian Bulletin of Fisheries and Aquatic Science 223. 148 pp.

Hewlett, J.D., and J.C. Fortson. 1983. Stream temperature under an inadequate buffer strip in the southern Piedmont. Water Resources Bulletin 18(6):983.

Hillman, G.R. 1987. Improving wetlands for forestry in Canada. Northern Forestry Centre, Canadian Forest Service, Inf. rep. Information Report NOR-X-288. 29 pp.

Hollingsworth, R.W. 1988. Fish habitat and forest fragmentation. Pages 19-25 in Convention of the Society of American Foresters. Rochester, N.Y.

Hooper, R.G. 1967. The influence of habitat disturbance on bird populations. M.S. thesis. Virginia Polytechnic Institute and State University, Blacksburg, Va.

Horwitz, E.L. 1978. Our nation's wetlands. U.S. Council on Environmental Quality, U.S. Government Printing Office, Washington, D.C. 70 pp.

House, R.A., and P.L. Boehne. 1987. The effects of stream cleaning on salmonid habitat and populations in a coastal Oregon drainage. Western Journal of Applied Forestry 2(3):84-87.

Howard, R.J., and J.A. Allen. 1988. Streamside habitats in southern forested wetlands: their role and implications for management. Pages 97-106 in The forested wetlands of the southern United States: proceedings of the symposium. 12-14 July 1988, Orlando, Fla.

Jackson, J.K., and V.H. Resh. 1989. Activities and ecological role of adult aquatic insects in the riparian zone of streams. Pages 342-345 in Proceedings of the California Riparian Systems Conference. 22-24 September 1988, Davis, Calif. USDA Forest Service General Technical Report PSW-110.

Jahn, L.R. 1978. Values of riparian habitats to natural ecosystems. Pages 157-160 in Strategies for protection and management of floodplain wetlands and other riparian ecosystems: proceedings of the symposium. 11-13 December 1978, Pine Mountain, Ga. USDA Forest Service General Technical Report WO-12.

Johnson, W.N. Jr., and P.W. Brown. 1990. Avian use of a lakeshore buffer strip and an undisturbed lakeshore in Maine. Northern Journal of Applied Forestry 7(3):114-117.

Kistritz, R.U., and G.L. Porter. 1993. Proposed wetland classification system for British Columbia: a discussion paper. B.C. Ministry of Forests, B.C. Ministry of Environment, Lands, and Parks, and B.C. Conservation Data Centre. 69 pp.

Knopf, F.L., and F.B. Samson. 1994. Scale perspectives on avian diversity in western riparian ecosystems. Conservation Biology 8(3):669-676.

Kostadinov, S.C., and S.S. Mitrovic. 1994. Effect of forest cover on the stream flow from small watersheds. Journal of Soil and Water Conservation 49(4):382-386.

Landin, M.C. 1979. The importance of wetlands in the North Central and Northeast United States to non-game birds. Pages 179-188 in R.M. DeGraaf and K.E. Evans (eds.). Management of north central and northeastern forests for nongame birds: workshop proceedings. 23-25 January 1979, Minneapolis, Minn. USDA Forest Service General Technical Report NC-51.

Lang, A.I. 1991. Status of the American coot, *Fulica americana*, in Canada. Canadian Field Naturalist 105(4):530-541.

Licht, L.E. 1971. The ecology of coexistence in two closely related species of frogs (*Rana*). Ph.D. thesis, University of British Columbia, Vancouver, B.C. 27 pp.

MacIntyre, D.H., and R.V. Palermo. 1980. The current status of the Amphibia in British Columbia. In R. Stace-Smith, L. Johns, and P. Joslin (eds.). Proceedings, symposium on threatened and endangered species and habitats in British Columbia and the Yukon. Fish and Wildlife Branch, B.C. Ministry of Environment, Lands and Parks, Victoria, B.C. 302 pp.

McComb, W., K. McGarigal, and R. Anthony. 1993. Small mammal and amphibian abundance in streamside and upslope habitats of mature Douglas-fir stands, western Oregon. Northwest Science 67(1):7-15.

McDade, M.H., F.J. Swanson, W.A. McKee, J.F. Franklin, and J. Van Sickle. 1990. Source distances for coarse woody debris entering small streams in western Oregon and Washington. Canadian Journal of Forest Research 20:326-330.

McGarigal, K., and W.C. McComb. 1992. Streamside versus upslope breeding bird communities in the central Oregon Coast Range. Journal of Wildlife Management 56(1):10-23.

McGurk, B.J. 1989. Predicting stream temperature after riparian vegetation removal. Pages 157-164 in Proceedings of the California Riparian Systems Conference, 22-24 /September 1988, Davis, Calif. USDA Forest Service General Technical Report PSW-110.

Meehan, W.R. 1991. Influences of forest and rangeland management on salmonid fishes and their habitats. American Fisheries Society Special Publication 19. 751 pp.

Meidinger, D., and J. Pojar. 1991. Ecosystems of British Columbia. B.C. Ministry of Forests, Victoria, B.C. 330 pp.

Melton, B.L., R.L. Hoover, R.L. Moore, and D.J. Pfankuch. 1983. Aquatic and riparian wildlife. Pages 262-301 in R.L. Hoover and D.L. Wills (eds.). Managing forested lands for wildlife. Colorado Division of Wildlife and USDA Forest Service RM, Denver, Colo.

Merendino, M.T., and C.D. Ankney. 1994. Habitat use by Mallards and American Black Ducks breeding in Central Ontario. Condor 96(2):411-421.

Miller, E. 1987. Effects of forest practices on relationships between riparian area and aquatic ecosystems. Pages 40-47 in Managing southern forests for wildlife and fish: a proceedings. January 1987, New Orleans, La. USDA Forest Service General Technical Report SO-65.

Murphy, M.L., and K.V. Koski. 1989. Input and depletion of woody debris in Alaska streams and implications for streamside management. North American Journal of Fisheries Management 9:427-436.

Myren, R.T., and R.J. Ellis. 1982. Evapotranspiration in forest succession and long-term effects upon fishery resources: a consideration for management of old-growth forests. Pages 183-186 in Fish and wildlife relationships in old-growth forests: proceedings of a symposium. April 1982, Juneau, Alaska.

Naiman, R.J., H. Decamps, and M. Pollock. 1993. The role of riparian corridors in maintaining regional biodiversity. Ecological Applications 3(2):209-212.

Narver, D.W. 1972. A survey of some possible effects of logging on two eastern Vancouver Island streams. Fisheries Research Board of Canada, Technical Report 323. 55 pp.

Newbold, J.D., D.C. Erman, and K.B. Roby. 1980. Effects of logging on macroinvertebrates in streams with and without buffer strips. Canadian Journal of Fisheries and Aquatic Sciences 37:1076-1085.

North American Wetlands Conservation Council (Canada). 1993. The Canadian wetland classification system. Canadian Wildlife Service and Environment Canada IP No. 93-3. Ottawa, Ont.

Nummi, P., H. Poysa, J. Elmberg, and K. Sjoberg. 1994. Habitat distribution of the Mallard in relation to vegetation structure, food, and population density. Hydrobiologia 279/280:247-252.

Nutter, W.L., and J.W. Gaskin. 1988. Role of streamside management zones in controlling discharges to wetlands. Pages 81-84 in The forested wetlands of the southern United States: proceedings of the symposium. 12-14 July 1988, Orlando, Fla. USDA Forest Service General Technical Report SE-50.

Oakley, A.L., J.A. Collins, L.B. Eversson, D.A. Heller, J.C. Howerton, and R.E. Vincent. 1985. Riparian zones and freshwater wetlands. Pages 57-80 in Management of wildlife and fish habitats in forests of western Oregon and Washington. USDA R6-F&WL-192-1985.

Orchard, S. 1990. Amphibians in B.C.: forestalling endangerment. Bioline 9(2):22-24.

Osborne, J.G. 1980. Effects of logging on resident and searun cutthroat trout (*Salmo clarki*) in small tributaries of Clearwater River, Jefferson County, Washington, 1978-1979. Final Report FRI-UW-8018. Fisheries Research Institute, University of Washington, Seattle. 56 pp.

Pojar, J., and H. Roemer. 1987. Wetland conservation and ecological reserves in British Columbia. Pages 617-624 in C.D.A. Rubec and R.P. Overend (eds.). Symposium '87: Wetlands/peatlands. 23-27 August 1987, Edmonton, Alta.

Potts, D.F., and B.K.M. Anderson. 1990. Organic debris and the management of small stream channels. Western Journal of Applied Forestry 5(1):25-28.

Ray, G.C. 1992. Coastal zone biodiversity patterns. BioScience 41(7):490-498.

Reiser, D.W., and T.C. Bjorn. 1979. Habitat requirements of anadromous salmonids. USDA Forest Service General Technical Report PNW-96. Portland, Ore. 54 pp.

Runka, G.G., and T. Lewis. 1981. Preliminary wetlands managers manual, Cariboo Resource Management Region. Assessment and Planning Division, B.C. Ministry of Environment. Technical Paper No. 5. 1st ed. Victoria, B.C. 113 pp.

Ryder, R.A. 1951. Waterfowl production in the San Luis Valley, Colorado. M.S. thesis, Colorado A & M College, Ft. Collins, Colo. 166 pp.

Savard, J.P., W.S. Boyd, and G.E.J. Smith. 1994. Waterfowl-wetland relationships in the Aspen Parkland of British Columbia: comparison of analytical methods. Hydrobiologia 279/280:309-325.

Sedell, J.R., P.A. Bisson, F.J. Swanson, and S.V. Gregory. 1988. What we know about large trees that fall into streams and rivers. Pages 47-81 in C. Maser, R.F. Tarrant, J.M. Trappe, and J.F. Franklin (eds.). From the forest to the sea: a story of fallen trees. USDA Forest Service General Technical Report PNW-GTR-229. Published in cooperation with the USDI Bureau of Land Management. Portland, Ore. 153 pp.

Shapley, S., and D.M. Bishop. 1965. Sedimentation in a salmon stream. Journal of Fisheries Research Board of Canada 22(4):919-928.

Sheridan, W.L., M.P. Perensovich, T. Faris, and K. Koski. 1982. Sediment content of streambed gravels in some pink salmon spawning streams in Alaska. Pages 153-165 in Fish and wildlife relationships in old-growth forests: proceedings of a symposium. April 1992. Juneau, Alaska.

Simons, L.H. 1985. Small mammal community structure in old growth and logged riparian- habitat. Pages 505-506 in North American Riparian Conference. 16-18 April 1985, University of Arizona, Tucson, Ariz.

Stevens, L.E., B.T. Brown, J.M. Simpson, and R.R. Johnson. 1977. The importance of riparian habitat to migrating birds. Pages 156- 164 in Importance, preservation and management of riparian habitat: a symposium. July 1997, Tuscon, Ariz. USDA Forest Service General Technical Report RM-43.

Stevens, V., F. Backhouse, and A. Eriksson. 1995. Riparian management in British Columbia: an important step towards maintaining biodiversity. Research Branch, B.C. Ministry of Forests; Habitat Protection Branch, B.C. Ministry of Environment, Lands and Parks, Victoria, B.C. Working Paper 13/1995. 30 pp.

Sustaining Wetlands Forum. 1990. Sustaining wetlands. Ottawa, Ont. 20 pp.

Swanson, F.J., S.V. Gregory, J.R. Sedell, and A.G. Campbell. 1982. Land-water interactions: the riparian zone. Pages 267-291 in Robert L. Edmonds (ed.). Analysis of coniferous forest ecosystems in the western United States. Hutchinson Ross Publishing, Stroudsburg, Pa.

Swift, L.W. Jr., and S.E. Baker. 1973. Lower water temperatures within a streamside buffer strip. USDA Forest Service RN SE-193. Asheville, N.C. 7 pp.

Tarnocai, C. 1980. Canadian wetland registry. Pages 9-39 in C.D.A. Rubec and F.C. Pollett (eds.). Proceedings, workshop on Canadian wetlands. Lands Directorate, Environment Canada. Ecological Land Classification Series, No. 12. Ottawa, Ont.

Terres, J.K. 1991. The Audubon Society encyclopedia of North American birds. Wing Books, New York. 1,109 pp.

Toews, D.A.A., and M.K. Moore. 1982. The effects of three streamside logging treatments on organic debris and channel morphology of Carnation Creek. Pages 129-153 in G. Hartman (ed.). Proceedings. Carnation Creek workshop: a ten-year review. 24-26 February 1982. Malaspina College, Nanaimo, B.C.

Triquet, A.M., G.A. McPeek, and W.C. McComb. 1990. Songbird diversity in clearcuts with and without a riparian buffer strip. Journal of Soil and Water Conservation 45(4):500-503.

Triska, F.J., J.R. Sedell, and S.V. Gregory. 1982. Coniferous forest streams. Pages 292-332 in R.L. Edmonds (ed.). Analysis of coniferous forest ecosystems in the western United States. Hutchinson Ross Publishing, Stroudsburg, Pa.

Tschaplinski, P.J. 1987. Comparative ecology of stream-dwelling and estuarine juvenile coho salmon (*Oncorhynchus kisutch*) in Carnation Creek, Vancouver Island, British Columbia. Ph.D. thesis. University of Victoria, B.C. xviii + 527 pp.

van Groenewould. 1977. Interim report for the use of buffer strips for the protection of small streams in the Maritimes. Canadian Forest Service, Inf. Rep. Information Report M-X-74. New Brunswick Department of Fisheries and Environment, Fredericton.

Van Ryswyk, A.L., K. Broesma, and J.W. Hall. 1992. Agricultural use and extent of British Columbia wetlands. Agriculture Canada, Research Branch, Kamloops, B.C. 129 pp.

Welsh, H.H. Jr., and A.J. Lind. 1991. The structure of the herpetofaunal assemblage in the Douglas-fir/hard wood forests of northwestern California and southwestern Oregon. Pages 395-413 in L.F. Ruggiero, K.B. Aubry, A.B. Carey, and M.H. Huff (eds.). Wildlife and vegetation of unmanaged Douglas-fir forests. USDA Forest Service General Technical Report PNW-GTR-285. Portland, Ore.

Wilcove, D., M. Bean, and P.C. Lee. 1992. Fisheries management and biological diversity: problems and opportunities. Pages 373-383 in R.E. McCabe (ed.). Transactions of the 57th North American Wildlife and Natural Resources Conference. Wildlife Management Institute, Washington, D.C.

Wilzbach, M.A. 1989. How tight is the linkage between trees and trout? Pages 250-255 in Proceedings of the California Riparian Systems Conference. 22-24 September 1988, Davis, Calif. USDA Forest Service General Technical Report PSW-110.

Yahner, R.H. 1988. Changes in wildlife communities near edges. Conservation Biology 2(4):333-339.

Zoltai, S.C., F.C. Pollett, J.K. Jeglum, and G.D. Adams. 1975. Developing a wetland classification for Canada. Pages 497-511 in B. Bernier and C.H. Winget (eds.). Proceedings, 4th North American Forest Soils Conference. Laval University Press, Quebec.

5
Interior Habitat
Angela von Sacken

Definitions

Interior habitat is generally defined as the portion of habitat that does not experience edge effects (see Chapter 8 for a discussion of edge effects), and thus maintains its functional viability for plant and animal communities (Laurance and Yensen 1991; Saunders et al. 1991; Chen et al. 1992; Forman 1995). Most research has been carried out on 'forest' interior habitat, but the concept applies equally to other interior habitats, such as forest meadow, grasslands, and agricultural fields.

The amount of interior habitat within a patch is determined by the size and shape of the patch and by the edge variable being measured. A forest patch must be above a minimum size and must be of a certain configuration in order to retain an element of interior habitat. If a patch falls below a minimum size or has a high perimeter/area ratio (such as a narrow elongated patch), it becomes permeable to edge influences. Interior habitat is dynamic; it can expand and contract through time depending on the surrounding matrix. Whether the interior habitat of a forest patch maintains its functional viability for plant and animal communities depends also on the plant or animal of concern. What constitutes interior habitat for one species may not for another, and must be considered in a species-specific context. As a general guideline, interior habitat has been documented as beginning anywhere from 20 m to 600 m from the edge (Chen et al. 1992).

It is important to note that interior forest can be found in forest of any age of sufficient size and configuration, whether the forest consists of coniferous, deciduous, or mixed stands. Much of the literature is limited to a discussion of old seral forests (coniferous and deciduous), because this subset of interior forest is the most difficult to maintain and manage. The term 'interior habitat' is also applicable to a forest opening because, at some distance, the edge effects of the adjacent mature timber no longer penetrate the interior of the opening (Angelstam 1992).

An example of a contiguous forest on the lower west coast of British Columbia (Photograph by Steve Voller)

Interior forest; interior forest habitat: interchangeable terms that describe one type of interior habitat.

Interior species: species that are area-sensitive and sensitive to isolation disturbance and edge effects (Alverson et al. 1994; Forman 1995).

Contiguous forest: an area of continuous forest that is uninterrupted by an induced forest edge such as that found between a cutblock and the forest; a term often used to describe extensive areas of old forest.

Patch: a relatively homogeneous unit of vegetation that is different from the matrix in which it occurs. Patches may occur at many scales: a canopy gap is a patch within a stand, and a stand is a patch within a landscape (McComb et al. 1993).

Gap: an opening produced by natural disturbance processes within a community. Gaps represent the microhabitats within a community (Forman 1995).

Opening: an area produced by major disturbances such as fire, windthrow, or clearcutting. Openings represent distinct ecosystems within the landscape (Forman 1995).

Reserve: a generic term used to define areas protected or managed to maintain their natural values. Reserves range in size from small to extremely large. Large reserves are a means of maximizing and maintaining interior habitat (Noss and Cooperrider 1994).

Background

What constitutes interior habitat is species- and microclimate- (wind versus solar radiation versus moisture, and so on) specific; there is therefore growing interest in understanding the role of interior habitat and its function in community dynamics. Interest has focused primarily on areas that were once dominated by contiguous habitat. Through forest harvesting, agriculture, and urban development, these once-large contiguous blocks have become fragmented, leaving behind isolated patches embedded within extensive areas of dissimilar habitat. The remnant patches are more susceptible to edge effects and are less likely to contain interior habitat. Most research on interior habitat has been carried out in late successional west coast (coastal Oregon, Washington, and British Columbia) coniferous forests and east coast deciduous forests.

Prior to large-scale harvesting, forests persisted over longer periods of time because large-scale natural disturbances occurred less frequently than the current cutting cycle (see Chapter 1 for a discussion of natural disturbances). The longevity and continuity of these forests resulted in important niches with which certain species were associated (Thomas et al. 1990). The rate and pattern of forest harvesting now determine the amount of late successional forest interior habitat. The balance between the rate of initiation and the rate of succession of patches determines whether a landscape is aggrading or degrading in terms of interior habitat. Harvesting removes a portion of the trees and subjects the remaining smaller forested patch to edge effects (Angelstam 1992; Noss and Cooperrider 1994).

Past harvesting practices that encouraged the creation of edge converted contiguous old forests into a complex mosaic of successional stages interspersed with patches of late successional forest. Timber harvesting in the Pacific Northwest imposed a high rate of disturbance through a 40-80-year rotation (McComb et al. 1993). At a landscape scale, a higher proportion of early seral stages prevailed because of the accelerated rate of disturbance from timber harvesting. This shorter disturbance interval did not allow late successional ecosystems, and thus late successional interior conditions, to develop, resulting in reduced habitat for species and processes associated with old seral interior habitat (McComb et al. 1993; Spies et al. 1994).

In a fragmented landscape with remnant old-forest patches, young forests moving through succession act as a buffer against edge effects and enhance the quality and quantity of the interior forest found within the remnants (Harris 1984; Alverson et al. 1994). How quickly forest succession mitigates edge effects appears to vary with forest type, site conditions, and specific species requirements. Saunders et al. (1991) questioned whether equilibrium of old seral forest interior conditions through regrowth could occur, and likened it to an idealized endpoint that was never really reached, much like climatic climax.

The scale of disturbances has an important bearing on interior habitat. Angelstam (1992) speculated that if clearcuts were sufficiently small, some species might perceive the area as a single forested landscape rather than a series of forested stands. Forman (1995) made a distinction between an opening and a gap disturbance: a gap was an internal spot or microhabitat within a community, whereas an opening was a distinct ecosystem at the landscape scale. He felt that gaps were unlikely to interrupt interior habitat because they were filled either through lateral extension or through the growth of the existing vegetation within the gap (Forman 1995). An opening, on the other hand, represented the interruption of forest interior habitat.

McComb et al. (1993) remained undecided about the effects of small-scale natural disturbances (such as root rot and localized windthrow) on late successional interior habitat. However, a study in Hoosier National Forest in Indiana found that alternative harvesting techniques, such as group selection, created more edge than even-aged management systems, and were therefore more detrimental to some forest interior bird species (Gustafson and Crow 1994). Spies et al. (1994) also noted that the use of alternative cutting patterns as a means of maintaining interior forest habitat needed further investigation.

Fewer studies have documented the interior habitat of early or mid-seral forests. However, Chen et al. (1992), in their study of microclimate effects, noted that a 10 ha patch of western hemlock seedlings (up to 10 cm tall) contained no interior habitat, whereas 45% of a 100 ha patch was considered to have interior habitat (microclimate not affected by edge influences). In a similar study of Douglas-fir seedlings 31-100 cm tall, only 18% of a 10 ha patch and 5% of a 100 ha patch were affected by edge habitat (Chen et al. 1992). Furthermore, Angelstam (1992) noted how the edge effects caused by the adjacent mature forest reduced the interior habitat attributes of a clearcut; wind speed into the clearcut was reduced by 50% to a distance of three times the mature forest's tree height. Similarly, hedgerows in agricultural landscapes modify evaporation by casting an influence some distance into the field (Angelstam 1992).

Ecological Principles

Interior Forest Microclimate

In a fragmented landscape, the habitat within a remnant late successional forest patch possesses different climatic conditions than a clearcut or other adjacent habitat. The interior habitat of a late successional (Douglas-fir) forest is generally characterized by a relatively stable environment that is cool, dark, humid, and windless (Hunter 1990; Chen et al. 1992). The amount of interior habitat available in a late successional forest patch depends on the adjacent habitat type and on the contrast between the two. The more pronounced the contrast (the harder the edge), the wider the band of edge influence and thus the smaller the area of interior habitat that remains.

For example, the contrast between a clearcut and a late successional forest is strong, and therefore the climatic differences between the two will be pronounced. The edge effect will carry more strongly and further into each habitat. Furthermore, because wind, solar radiation, and moisture (the edge effects) will penetrate a stand to varying degrees, the amount of interior habitat will vary with each variable measured. Forest type (Douglas-fir versus ponderosa pine, deciduous versus conifer), forest structure, forest regrowth, topography, and weather patterns may also determine the amount of interior habitat available. Thus, because environmental conditions tend to vary along a continuum, there is no clearly defined line between interior and edge habitat (Alverson et al. 1994).

Wind speed and reach will vary as wind encounters different habitat types. As wind moves across an opening and comes into contact with a forest patch, the upper portion of the wind profile continues unimpeded as it moves above the forest crown. The lower portion, however, encounters the branches and stems of the forest vegetation and slows down, reaching equilibrium some distance inside the forest (Saunders et al. 1991; Fraver 1994). As wind reaches equilibrium within the forest patch, it may result in seed deposits of weed species (dandelion, burdock, selfheal, motherwort, plantain species, and thistle) along the gradient from edge to interior, which may eventually change the floristic composition of the interior habitat within the forest patch (Alverson et al. 1994).

Plant desiccation resulting from wind penetrating a forest patch may adversely affect species that require interior habitat. A Swedish study showed that arboreal lichens were adversely affected by both windfelling and desiccation, but noted that plant desiccation was most notable in the first year after creation of a forest opening (Angelstam 1992). Windthrow can further reduce the amount of interior habitat by physically reducing the size of the forest patch. The transfer of insects and disease organisms into a forest patch as a result of wind may also compromise interior habitat. Increased litter fall (leaves and stems) can alter soil surface characteristics and nutrient cycling, affecting ground-dwelling fauna (Franklin and Spies 1991; Saunders et al. 1991; Fraver 1994). The effect of wind on plant gas exchange, which affects photosynthesis, may have implications for interior habitat but has not been examined in any detail (Saunders et al. 1991).

The amount of solar radiation varies from edge to interior. The interior habitat of a late successional forest patch reflects much of the radiation, maintaining relatively constant cool forest interior temperatures. Shade-tolerant plant species have adapted to the moderate climatic conditions of forest interiors, and are unable to survive the temperature fluctuations in forest openings. On the other hand, some aggressive and weedy species are better adapted to temperature extremes and thus can thrive in forest openings. If such conditions (i.e., more openings) are continually created, these weedy species will eventually dominate the seed pool and, through seed rain, will alter the species composition within the remnant forest patches. The change in solar radiation balance brought about by opening the forest exposes interior-dwelling plants to a higher incidence of frost. In general, the interior of a clearcut has a higher daytime (soil and air) temperature and a lower nighttime temperature relative to the forest interior (Saunders et al. 1991). A secondary effect of increased solar radiation is the creation of thermal gradients between vegetation types, which result in turbulent air flow. This can lead to an influx of hot, dry air into the forest interior, which reduces humidity and increases evapotranspiration.

Moisture also varies along the gradient from edge to interior. The interior of an old seral forest has less snow accumulation but retains a snowpack

The interior of old-growth forests has less snow accumulation but retains a snowpack longer. (Photograph by Joan Voller)

longer. This results in a steady rate of moisture absorption, unlike the irregular moisture fluxes found in a forest opening. Furthermore, in coastal forests, the interior of old forest typically has multilayered canopies and high leaf area that are effective in condensing and precipitating moisture and atmospheric particles (Franklin and Spies 1991). In a study in the northern Oregon Cascade Range, fog drip in old seral forests was found to contribute 30% of the measurable precipitation each year (Franklin and Spies 1991). The relatively constant state of the hydrological cycle results in constant but low amounts of nutrient leaching and low rates of soil erosion. The effect is an overall higher level of water quality in large watershed-size forest interior patches (Franklin and Spies 1991).

Microclimatic edge effects may eventually lead to changes in the ecology of old seral forest patches, most notably through changes in plant abundance. Angelstam (1992) noted that species richness showed little change whereas species abundance increased markedly in response to changes in microclimate. Over a four-year period, annuals such as meadow cow wheat (*Melampyrum pratensese*) and woodland cow wheat (*Melampyrum sylvaticum*) spread from the edge of a black spruce remnant patch to the interior. Over the same period, the orchid heart-leaved twayblade (*Listera cordata*) showed a reduction in shoot numbers within remnant patches. It must be noted, however, that this study defined remnant patches as being up to 1 ha in size, indicating a great degree of permeability to edge effects. Angelstam (1992) discussed another example where there was a noted difference in the abundance of the perennial herb maiden pink (*Dianthus deltoidles*) between an unfragmented forest patch and a fragmented forest patch, resulting from fewer visits from insect pollinators within the fragmented patch. Franklin and Forman (1987) noted that the lichen *Lobaria oregana* depended strictly on late seral interior forest conditions, and was therefore susceptible to changes in microclimate.

Changes in microclimate may also directly affect fauna. Saunders et al. (1991) referred to a study that indicated that the feeding behaviour of ants may be altered through increased solar radiation. The study documented how the dominant ant species *Iridomyrmex* foraged only when temperatures were high and other ant species foraged only when *Iridomyrmex* was absent. Increased solar radiation would therefore create more foraging opportunities for *Iridomyrmex*, to the detriment of the other species. Another example involves the forest pest Vienna moth (*Ocneria monacha*), which lays its eggs on tree trunks. Trees at the edge of a patch receive higher levels of solar radiation than those in the interior of the patch, allowing the larvae to hatch earlier than their parasite, which emerges from the cooler forest floor. This gives the pest a head start and can unbalance the cycle (Geiger 1965).

Interior Habitat Requirements of Forest Dwellers

Forest-harvesting techniques that result in forest fragmentation (see Chapter 2) diminish the quantity and quality of remaining forest interior habitat. This is of greater concern in late successional forest, since this type of interior habitat is the most difficult to maintain and manage. Diminishing quantity refers to the lack of adequately sized and distributed late successional forest patches, which can impede migration and/or dispersal. Diminishing quality refers to the permeability of the remaining patches to predators, competitors, and climatic conditions such as wind and snowfall.

A number of species are termed 'old growth-associated,' but it is uncertain whether the specific structural attributes (e.g., snags and coarse woody debris) of late successional forests or the contiguity of the forests are more important to wildlife (FEMAT 1993). The Northern Spotted Owl (*Strix occidentalis*) is a good example. It finds optimal habitat in large patches of late successional forest, but it can also make use of large patches of younger successional forest in which old-forest structure has been retained (Thomas et al. 1990). Marten (*Martes americana*) has also been identified as an 'old growth-associated' species, and, much like the Northern Spotted Owl, prefers large patches of late successional forest. When this habitat is unavailable, martens make use of younger seral stages, selecting stands with higher levels of coarse woody debris (Lofroth 1993). It would appear that old-forest structure and adequately sized forest patches are inextricably linked.

Relatively few studies have documented changes in abundance of forest birds, small mammals, and amphibians as a result of forest fragmentation in western North America (Lehmkuhl and Ruggiero 1991). This is due partly to the lack of research on this specific topic and partly to the difficulty in quantifying research results for forested landscapes that are dynamic. Alternatively, extensive research from eastern North America has clearly noted a decline in eastern songbirds as patch size decreases and patch isolation increases. The difference in results between eastern and western studies may be attributed to the relatively static habitat boundaries in eastern forests. These forests tend to be bound by agriculture and urban developments, compared with dynamic boundaries resulting from forest regrowth in western coniferous forests (Lehmkuhl and Ruggiero 1991). Through forest regrowth, previously harvested patches act as a buffer against edge effects and enhance the quantity and quality of the interior habitat found in the remaining old-forest patches (Harris 1984; Alverson et al. 1994). Furthermore, many species have evolved in naturally heterogeneous landscapes (as a result of natural disturbance) that include edge habitat; interior-dependent species may therefore be a misnomer in some forested landscapes (Schieck et al. 1995).

Interior Habitat and Herpetofauna

Cool, moist conditions such as those found in old-forest interior habitat appear to be an important habitat requirement for most herpetofauna. Terrestrial amphibians exhibit a high microhabitat specificity (cool, moist habitats) and therefore a clear pattern of abundance in old and mature forests (Dupuis et al. 1995). In a study in the southern Appalachians, Petranka et al. (1993) found that salamanders were sensitive to changes in temperature, humidity, and soil moisture regime. Because salamanders lack lungs and exchange gases through cutaneous respiration, their skin must be kept moist. This is facilitated by moist forest floor microhabitats. Eliminating shade, reducing leaf litter, increasing soil surface temperature, and reducing soil surface moisture all negatively affect salamanders (Blymer and McGinnes 1977; Bury 1983; Petranka et al. 1993).

In the Cascade Range, Aubry and Hall (1991) found that, in general, older Douglas-fir stands with cooler temperatures and relatively flat slopes had the most abundant amphibian populations. However, they did not find substantially lower amphibian species richness or abundance in younger stands and attributed this to the high retention of old-forest structural attributes, such as coarse woody debris, in these stands. On Vancouver Island, Dupuis et al. (1995) found a significant difference in salamander abundance between old-growth forests and younger forests. They attributed this difference to the specific microclimatic conditions found in old-growth forests. Because of the relatively small area requirements of most amphibians, it is speculated that they may find adequate interior conditions in smaller forest patches than those required by larger vertebrates. For example, riparian reserve areas may be perceived as having interior habitat for terrestrial amphibians, but may not meet the needs of a larger vertebrate.

The point at which climatic conditions suitable for terrestrial amphibians occur in young forests is not well understood, but likely varies with ecosystem. Petranka et al. (1993) speculated that selective harvesting would maintain forest conditions suitable for salamanders but would require cutting more acreage. How this trade-off affects salamander abundance is not known.

Interior Habitat and Birds

The interior habitat requirements of eastern songbirds has been well documented. Interior forest-dwelling birds are species that require relatively large contiguous tracts of forest to support viable breeding populations. They are area-sensitive species and many require more than 150 ha of contiguous tracts of forest for successful breeding (Therres 1993). In general, insectivores are more area-sensitive than seed eaters or omnivores, and long-distance migrants are more area-sensitive than short-distance migrants or resident birds (Forman 1995). Forest fragmentation interrupts the contiguity and availability of forest habitat, making specific microhabitats such as streams, ponds, and steep slopes inaccessible to some species (Blake and Karr 1984).

Fragmentation leads to improved habitat for generalist species, and therefore to a higher incidence of these species. Generalist species compete with habitat specialists and/or act as potential predators, which eventually leads to a change in the faunal community (Yahner and Scott 1988; Yahner et al. 1989; Keller and Anderson 1992). As the forest patch shrinks and the edge-to-interior ratio increases, the incidence of brood parasitism is likely to increase as nest predators permeate the patch (Wilcove 1987; Andren and Angelstam 1988). Wilcove (1987) suggested that edge-related predation may extend as far as 600 m into the forest patch. Corvids and small mammals are the most serious nest predators. Forest fragmentation resulting from forest harvesting and urban and agricultural development has led to an overall reduction in the breeding populations of neotropical migrants in eastern North America. The same effect is not clearly documented in western North America, however.

In a study on Vancouver Island, British Columbia, Schieck et al. (1995) evaluated the species richness and abundance of old-growth bird species in old-growth remnants of various sizes. Remnants ranged from 4 ha to 500 ha in the Coastal Douglas-Fir Zone and from 20 ha to 2,500 ha in the Coastal Western Hemlock Zone. All remnants were more than 300 years old. This study indicated that patch size did not affect the abundance and richness of old-growth bird species. There are three possible explanations for these results: (1) abundance of old-growth species may not be affected, but population viability over the long run is; (2) because of the natural heterogeneity of these forests, old-growth species have coevolved with species dependent on other habitats, such as edge habitat; or (3) the effect on old-growth bird species in the smallest remnant patches may have been of little consequence because there was not much contrast between the old-forest remnant and the matrix. Similarly, Keller and Anderson (1992) suggested that it was the overall loss of habitat (diminishing quantity) rather than the edge effects (diminishing quality) that was detrimental to certain avian species in western North America. This may be due to the relatively early stage of forest fragmentation in western forests compared with eastern forests (Keller and Anderson 1992).

Manuwal (1991) suggested that coastal Douglas-fir forests have a higher proportion of resident birds than neotropical migrants, and therefore suitable wintering habitat that provides roosting sites and thermal cover is critical. Huff and Raley (1991) noted that late successional stands in general provide better winter habitat for most bird species than do younger stands. Manuwal (1991) suggested that a reduction in forest patch size

Caribou prefer the interior conditions of a mature or old-growth forest to mitigate the more extreme climatic conditions found in forest openings. (Photograph by Scott Harrison)

could reduce the winter carrying capacity for resident birds. In the Pacific Northwest, preliminary research results indicate that Winter Wren (*Troglodytes troglodytes*), Swainson's Thrush (*Catharus ustulatus*), and Varied Thrush (*Ixoreus naevius*) may be more abundant in the interior of forest patches (Lehmkuhl and Ruggiero 1991). Many years of data regarding reproductive success and the subsequent survival of offspring are probably needed to ascertain whether late successional interior forest patches are essential for the survival of avian species.

Interior Habitat and Mammals
Some large mammals, such as woodland caribou (*Rangifer tarandus*), utilize the microclimatic conditions found in interior forest habitat. Mature and old-growth forest patches offer thermal cover and mitigate the more extreme climatic conditions found in forest openings. The canopy of the late successional forest interior intercepts much of the snowfall, permitting easier winter feeding (Nyberg and Janz 1990; Lehmkuhl and Ruggiero 1991). These forests may also have a greater abundance of arboreal lichens, which make up a significant portion of the winter diet.

Fishers (*Martes pennanti*) are more abundant in closed-canopy forests and dispersing individuals appear to avoid non-forested areas, implying a need for habitat continuity (Powell and Zielinski 1994). Overhead cover for protection from avian predators is important to martens, which therefore select habitat with higher canopy closure. Lofroth (1993) noted that martens rarely ventured into recent clearcuts and avoided early seral forests. More specific information regarding interior forest habitat requirements is lacking (Thompson 1991; Buskirk and Ruggiero 1994).

Small, less mobile vertebrate and invertebrate species may be more susceptible to changes in interior conditions. It would appear that opening a previously closed canopy of continuous forest subjects such species to hostile climatic conditions and predation. Alternatively, large, wide-ranging species such as elk (*Cervus elaphus*) may find suitable interior habitat conditions in neighbouring watersheds (Bunnell and Chan-McLeod 1997).

Reserve (Patch) Design and Interior Forest Habitat
The size and shape of a forest patch or forest reserve influences the amount of interior habitat available. Interior forest habitat declines sharply when patches fall below some critical size threshold. To maximize habitat for interior species, a forest patch should be circular, continuous, and as large as possible (Laurance and Yensen 1991; Faaborg et al. 1993). Compact forms are effective in conserving resources. They protect internal resources against the detrimental effects of the surroundings, whereas convoluted or elongated forms are effective in enhancing interaction with the surroundings (Forman 1995). Many existing reserves or forest patches encompass large areas, but are irregular in shape and therefore have a small core of interior habitat. Fragments that are irregular in shape accrue edge effects more rapidly than circular fragments (Laurance and Yensen 1991). However, linear forest patches should not be discounted, especially riparian areas that are wide enough to supply both connectivity and interior habitat (Mladenoff et al. 1994). In order to increase the interior habitat of existing irregularly shaped or small forest patches, additional adjacent areas should be managed as part of such patches, increasing their circularity over the long term. The intent is to eliminate sharp ecotone boundaries and thus reduce edge effects (Laurance and Yensen 1991). Alternative methods advocated in the Mount Hood National Forest Plan are either to harvest small isolated blocks within the landscape while not harvesting large intact blocks, or to relegate harvesting to the margins of a larger block. Maintaining and managing interior habitat is closely linked with the spatial patterns of forested landscapes (see Chapter 2 for further discussion of spatial patterns).

At a landscape scale (within a managed landscape), interior habitat should be managed as a shifting mosaic where large patches of late successional forest containing interior habitat will appear and disappear over time while the total area of forest interior remains in a steady state. At a stand level, we do not yet know the scale of forest remnant required to meet the needs of individual species.

Research Needs

Areas for further research include the following:

Breeding bird studies. In the Pacific Northwest, long-term breeding bird studies are needed to note breeding success and subsequent survival within forest patches of various sizes and seral stages. Are late successional forest patches of adequate size essential to the long-term survival of certain breeding birds (Manuwal 1991)?

Habitat uses. There is a need to better understand how climatic and seasonal conditions influence the interior habitat requirements of fauna in the Pacific Northwest. Is there a greater need for interior forest conditions during the winter for thermal cover and feeding requirements (Huff and Raley 1991)?

Definition. There is a need to understand whether there are 'interior-dependent' species in the Pacific Northwest or whether certain species are just dependent on late seral structural attributes. More research is required on species and species groups associated with late successional forests (Carey et al. 1991; Lehmkuhl and Ruggiero 1991).

The effects of forest fragmentation. More research is needed into the processes affecting wildlife (birds, mammals, reptiles, and amphibians) persistence in fragmented landscapes. The consequences of fragmentation over space and time in the Pacific Northwest must be examined.

Edge influence through time. A better understanding of how secondary succession affects the interior habitat in remnant patches is needed. How quickly does forest regrowth mitigate edge effects, and do individual tree species mitigate edge effects more quickly than others? (Angelstam 1992)?

Comparison of managed and natural forest interior conditions. Comparative studies are needed on species use and preference for natural (naturally regenerating from fire or windthrow) forest interior habitat versus managed (planted) forest interior habitat.

Maintaining interior habitat through alternative silvicultural systems. What is the effect of alternative silvicultural systems (strip-cutting, patch-cutting, individual tree selection, and so on) on interior habitat? Are such systems a means of harvesting some of the wood while maintaining forest interior habitat for species that require it (Angelstam 1992; McComb et al. 1993; Gustafson and Crow 1994; Spies et al. 1994)?

Small-scale natural disturbances. How do individual tree mortality and other small-scale natural disturbances affect forest interior habitat (McComb et al. 1993)?

Plant gas exchange. More research is needed on the effect of increased wind on plant gas exchange. An increase in wind can lead to a decrease in plant photosynthesis. Will the microclimatic effects of wind result in decreased plant productivity (Saunders et al. 1991)?

Fawn lilies such as those shown above are commonly seen in the old growth along the Cowichan River on Vancouver Island. Further research into which plant species may require interior conditions should be undertaken. (Photograph by Steve Voller)

Interior habitat in British Columbia. Much of the literature on interior habitat involves studies carried out in coastal Oregon and Washington. Studies of interior habitat should be initiated in central and northern British Columbia.

Vascular and nonvascular plants and interior habitat. What plant species are dependent on interior habitat?

Literature Cited

Alverson, W.S., W. Kuhlmann, and D.M. Waller. 1994. Wild forests: conservation biology and public forestry. Island Press, Washington, D.C. 300 pp.

Andren, H., and P. Angelstam. 1988. Elevated predation rates as an edge effect in habitat islands: experimental evidence. Ecology 69(2):544-547.

Angelstam, P. 1992. Conservation of communities – the importance of edges, surroundings, and landscape mosaic structure. Pages 9-70 in Lennart Hansson (ed.). Ecological principles of nature conservation. Elsevier Applied Science, New York.

Aubry, K.B., and P.A. Hall. 1991. Terrestrial amphibian communities in the southern Washington Cascade Range. Pages 353-362 in L.F. Ruggiero, K.B. Aubry, A.B. Carey, and M.H. Huff (eds.). Wildlife and vegetation of unmanaged Douglas-fir forests. USDA Forest Service General Technical Report PNW-GTR-285. Portland, Ore.

Blake, J.G., and J.R. Karr. 1984. Species composition of bird communities and conservation benefit of large versus small forests. Biological Conservation 30:173-187.

Blymer, M.J., and B.S. McGinnes. 1977. Observations on possible detrimental effects of clearcutting on terrestrial amphibians. Bulletin of the Maryland Herpetological Society 13(2):79-83.

Bunnell, F., and A. Chan-McLeod. 1997. Terrestrial vertebrates. Pages 103-130 in Schoonmaker, P., B. von Hagen, and E.C. Wolf (eds.). The rain forests of home: profile of a North American bioregion. Island Press, Washington, D.C.

Bury, B.R. 1983. Differences in amphibian populations in logged and old-growth redwood forest. Northwest Science 57(3):167-178.

Buskirk, S.W., and L.F. Ruggiero. 1994. American marten, fisher, lynx and wolverine in the Western United States. USDA Forest Service General Technical Report RM-254:7-30.

Carey, A.B., M.M. Hardt, S.P. Horton, and B.L. Biswell. 1991. Spring bird communities in the Oregon Coast Range. Pages 123-144 in L.F. Ruggiero, K.B. Aubry, A.B. Carey, and M.H. Huff (eds.). Wildlife and vegetation of unmanaged Douglas-fir forests. USDA Forest Service General Technical Report PNW-GTR-285. Portland, Ore.

Chen, J., J.F. Franklin, and T.A. Spies. 1992. Vegetation responses to edge environments in old-growth Douglas-fir forests. Ecological Applications 2(4):387-396.

Dupuis, L., A. James, N.M. Smith, and F. Bunnell. 1995. Relation of terrestrial-breeding amphibian abundance to tree-stand age. Conservation Biology 9(3):645-653.

Faaborg, J., M. Brittingham, T. Donovan, and J. Blake. 1993. Habitat fragmentation in the temperate zone: a perspective for managers. Pages 331-338 in D. Finch and P.W. Stangle (eds.). Status and management of neotropical migratory birds. USDA Forest Service General Technical Report RM-229.

Forest Ecosystem Management Assessment Team (FEMAT). 1993. Forest ecosystem management: an ecological, economic and social assessment. USDA Forest Service (and other agencies), Portland, Ore.

Forman, R.T. 1995. Land mosaics: the ecology of landscapes and region. Cambridge University Press, Cambridge, England. 632 pp.

Franklin, J.F., and R.T. Forman. 1987. Creating landscape patterns by forest cutting: ecological consequences and principles. Landscape Ecology 1(1):5-18.

Franklin, J.F., and T. Spies. 1991. Composition, function, and structure of old-growth Douglas-fir forests. Pages 71-80 in L.F. Ruggiero, K.B. Aubry, A.B. Carey, and M.H. Huff (eds.). Wildlife and vegetation of unmanaged Douglas-fir forests. USDA Forest Service General Technical Report PNW-GTR-285. Portland, Ore.

Fraver, S. 1994. Vegetation responses along edge to interior gradients in the mixed hardwood forests of the Roanoke River Basin, North Carolina. Conservation Biology 8(3):822-832.

Geiger, R. 1965. The climate near the ground. Harvard University Press, Cambridge, Mass. 611 pp.

Gustafson, J.E., and T.R. Crow. 1994. Forest management alternatives in the Hoosier National Forest. Journal of Forestry (August):28-29.

Harris, L.D. 1984. The fragmented forest. University of Chicago Press, Chicago, Ill. 211 pp.

Huff, M.H., and C.M. Raley. 1991. Regional patterns of diurnal breeding bird communities in Oregon and Washington. Pages 177-205 in L.F. Ruggiero, K.B. Aubry, A.B. Carey, and M.H. Huff (eds.). Wildlife and vegetation of unmanaged Douglas-fir forests. USDA Forest Service General Technical Report PNW-GTR-285. Portland, Ore.

Hunter, M.L. Jr. 1990. Wildlife and forests and forestry: principles of managing forests for biological diversity. Prentice Hall, Englewood Cliffs, N.J. 370 pp.

Keller, M.E., and S.H. Anderson. 1992. Avian use of habitat configurations created by forest cutting in southeastern Wyoming. Condor 94:55-65.

Laurance, W.F., and E. Yensen. 1991. Predicting the impacts of edge effects in fragmented habitats. Biological Conservation 55:77-92.

Lehmkuhl, J.F., and L.F. Ruggiero. 1991. Forest fragmentation in the Pacific Northwest and its potential effects on wildlife. Pages 35-46 in L.F. Ruggiero, K.B. Aubry, A.B. Carey, and M.H. Huff (eds.). Wildlife and vegetation of unmanaged Douglas-fir forests. USDA Forest Service General Technical Report PNW-GTR-285. Portland, Ore.

Lofroth, E.C. 1993. Scale dependent analysis of habitat selection by marten in the subboreal spruce biogeoclimatic zone. M.Sc. thesis, Simon Fraser University, Burnaby, B.C. 109 pp.

McComb, W.C., T.A. Spies, and W.H. Emmingham. 1993. Douglas-fir forests: managing for timber and mature habitat. Journal of Forestry (December):31-42.

Manuwal, D.A. 1991. Spring bird communities in the southern Washington Cascade Range. Pages 207-220 in L.F. Ruggiero, K.B. Aubry, A.B. Carey, and M.H. Huff (eds.). Wildlife and vegetation of unmanaged Douglas-fir forests. USDA Forest Service General Technical Report PNW-GTR-285. Portland, Ore.

Mladenoff, D.J., M. White, T. Crow, and J. Pastor. 1994. Applying principles of landscape design and management to integrate old-growth forest enhancement and commodity use. Conservation Biology 8(3):752-762.

Noss, R.F., and A.Y. Cooperrider. 1994. Saving nature's legacy: protecting and restoring biodiversity. Island Press, Washington, D.C. 416 pp.

Nyberg, J.B., and D.W. Janz (eds.). 1990. Deer and elk habitats in coastal forests of southern British Columbia. B.C. Ministry of Forests Special Report Series No. 5. 310 pp.

Petranka, J.W., M.E. Eldridge, and K.E. Haley. 1993. Effects of timber harvesting on Southern Appalachian salamanders. Conservation Biology 7(2):363-369.

Powell, R.A., and W.J. Zielinski. 1994. Fisher. Pages 38-66 in L.F. Ruggiero, K.B. Aubry, S.W. Buskirk, L.J. Lyon, and W.J. Zielinski (eds.). American marten, fisher, lynx and wolverine in the western United States. USDA Forest Service General Technical Report RM-254.

Saunders, D.A., R.J. Hobbs, and C.R. Margules. 1991. Biological consequences of ecosystem fragmentation: a review. Conservation Biology 5(1):18-32.

Schieck, J., K. Lertzman, B. Nyberg, and R. Page. 1995. Effect of patch size on birds in old-growth montane forests. Conservation Biology 9(5):1072-1084.

Spies, T.A., W. Ripple, and G.A. Bradshaw. 1994. Dynamics and patterns of a managed coniferous forest landscape in Oregon. Ecological Applications 4(3):555-568.

Therres, G.D. 1993. Integrating management of forest interior migratory birds with game in the Northeast. Pages 402-407 in D. Finch and P.W. Stangle (eds.). Status and management of neotropical migratory birds. USDA Forest Service General Technical Report RM-229.

Thomas, J.W., E.D. Forsman, J.B. Lint, E.C. Meslow, B.R. Noon, and J. Verner. 1990. A conservation strategy for the Northern Spotted Owl. Interagency Scientific Committee to Address the Conservation of the Northern Spotted Owl, Portland Ore. 458 pp.

Thompson, I.D. 1991. Could marten become the Spotted Owl of Eastern Canada? The Forestry Chronicle 67(2):136-140.

Wilcove, D.S. 1987. From fragmentation to extinction. Natural Areas Journal 7(1):23-29.

Yahner, R.H., and D.P. Scott. 1988. Effects of forest fragmentation on depredation of artificial nests. Journal of Wildlife Management 52(1):158-161.

Yahner, R. H., T.E. Morrell, and J.S. Rachael. 1989. Effects of edge contrast on depredation of artificial avian nests. Journal of Wildlife Management 53(4):1135-1138.

6
Biodiversity and Old-Growth Forests
Andy MacKinnon

Definitions

Old-growth forest: Old-growth forests have been defined in different ways for different purposes. Pojar et al. (1992) and B.C. Ministry of Forests (1992) characterize British Columbia's old-growth forests as having the following attributes:

- relatively old (older than the average catastrophic disturbance interval)
- composed of trees older than the average life span for the species
- more diverse (structurally and biologically) than young stands
- undisturbed by humans.

Definitions of old-growth forests applicable to British Columbia can be found in B.C. Ministry of Forests (1992).

Old growth-associated species: The concept of association is a broad one – that a species is usually found in a particular habitat. Ruggiero et al. (1991) considered species to be 'associated' when they were numerically more abundant in old growth compared with other seral stages, or required habitat features characteristic of old growth. They considered species to be 'closely associated' when they were significantly more abundant, based on levels of statistical significance set by each investigator, in old growth compared with other seral stages. Carey's (1989) concept of 'closely associated' is similar.

Old growth-dependent species: The concept of dependency is more restrictive, implying that a species requires old-growth forests for its survival. Ruggiero et al. (1988) define ecological dependency as 'the relationship between a population and the environment(s) required for its persistence.'

Biodiversity: The diversity of plants, animals, and other living organisms in all their forms and levels of organization, including the diversity of genes, species, and ecosystems, as well as the evolutionary and functional processes that link them.

Much of British Columbia's biodiversity is associated with old-growth forests. Wells Gray and many other B.C. parks hold much of the province's pristine old growth. (Photograph by Joan Voller)

Background

Over 60% of British Columbia is forested. Large portions of the province (mostly in wetter climatic zones) have been intermittently covered with old-growth forests for thousands of years. There is great variation, structurally and biologically, from one ecological zone of the province to another (Pojar et al. 1992). It is not surprising that much of B.C.'s biodiversity is associated with old-growth forests.

The area of old-growth forest remaining in some parts of the province (especially southwestern B.C.) has been greatly reduced by human activities. There is essentially no first-growth forest left in the Coastal Douglas-Fir (CDF) biogeoclimatic zone (MacKinnon and Eng 1995) or the lower Fraser Valley. Second-growth forests provide habitat for some, but not all, of the organisms that were original inhabitants of old-growth forests.

To recognize which species are associated with old-growth forests, and the nature of that association, is therefore important in managing for biodiversity. It may be possible to provide some attributes of old-growth forests in managed second-growth forests. Species not restricted to old-growth forests may require attributes of old growth that will be difficult, or impossible, to provide in second-growth forests. For most species that live in old-growth forests, the nature of their association with such forests – whether they occur there incidentally or are obligately dependent – remains unknown.

Biodiversity describes organization at a variety of levels, including ecosystems, species, and genomes. This chapter deals primarily with species diversity in old-growth forests. At the same time, it should be kept in mind that in a forested province such as B.C., maintaining ecosystem diversity implies maintaining some undisturbed samples of old growth. Furthermore, if the range of an old growth-associated species is reduced, we can assume that the genetic diversity of that species has also been reduced.

This chapter considers old-growth habitat primarily at the stand level. However, old-growth forests also provide important habitat at a landscape level – in corridors, interior areas, and riparian zones – in ways that are discussed in detail in other chapters. Information is presented about the old-growth associations of organisms whose ranges include B.C.; some of the research took place in B.C. and some in adjacent jurisdictions.

Managing for biodiversity in B.C. requires managing for a variety of seral habitats, including (but certainly not restricted to) old growth. The role of B.C.'s old-growth forests in maintaining biodiversity is discussed by Nyberg et al. (1987), Fenger and Harcombe (1990), B.C. Ministry of Forests (1992), Pojar et al. (1992), and Harding (1994).

Association and Dependence

The concept of old-growth association and dependence has been discussed by a number of authors (e.g., Carey 1984; Ruggiero and Carey 1984; Ruggiero et al. 1988; Carey 1989; Ruggiero 1991; FEMAT 1993). Various reports (Carey 1989; Fenger and Harcombe 1990; Ruggiero et al. 1991; FEMAT 1993; Thomas et al. 1993; Craig, in prep.) have produced lists of species described as associated with, or dependent on, old-growth forests.

The concept of association is a broad one, i.e., that a species is usually found in a particular habitat. The list of 1,105 species 'closely associated with late successional forests in National Forests within the range of the Northern Spotted Owl' (FEMAT 1993, based in part on Thomas et al. 1993) is an example of a list of species associated with late successional (including old-growth) forests. The 1,105 species include 106 bryophytes, 527 fungi, 157 lichens, 124 vascular plants, 102 molluscs, 18 amphibians, 7 fish (races/species/groups), 38 birds, and 26 mammals.

The concept of dependency is more restrictive, implying that a species requires old-growth forests for its survival. But as Ruggiero (1991) and other authors point out, determining whether or not a species has an absolute requirement for old-growth forests for its survival is not an easy task:

> Nevertheless, a close association with a habitat should be interpreted to indicate ecological requirements for persistence. This condition must be accepted as indicating dependency unless more intensive research supports a different interpretation. Equivocating and insisting on 'absolute proof' of dependency before committing to the appropriate management activities is inappropriate. The scientists who provide research results and the managers who use those results must recognize that such absolute knowledge is usually not attainable.

The lists of old growth-associated species provided here for various taxonomic groups include species that may or may not be obligately dependent on old-growth forests. In addition, it should be kept in mind that species that depend on old-growth forests in one part of the province might not have that dependence in another.

Species diversity is but one component of biodiversity. Raphael (1991), for example, found that in unmanaged Douglas-fir stands in western Washington and Oregon, total vertebrate species diversity varied little between age classes. His conclusion:

> Landscapes composed entirely of clearcuts, old growth, young forest, or any combination of ages would support similar numbers of vertebrate species ... Do these results suggest that old growth could be eliminated without negative effects on biodiversity? The answer is yes, but only if the number of species is all that counts ... a substantial proportion of species are significantly more abundant in certain age classes. Loss of preferred age classes may reduce the population viability of these species. Species richness is only one component of biodiversity.

Ecological Principles

Maintaining habitat for old growth-associated species requires a clear understanding of the nature of the association. Some habitat features (generally referred to as 'attributes') of old-growth forests can be managed for in second-growth forests; examples include snags and coarse woody debris (CWD). Other attributes, such as large-diameter snags, may not be so easy to manage for. (See Chapter 7 for more information on CWD and wildlife trees.) As well, some species may require aspects of old growth (such as maintenance of interior habitat) incompatible with any forest development.

Bunnell and Kremsater (1990) suggest three reasons why management for old growth-associated species is important:

- old growth (or its attributes) are not created easily or quickly
- adaptation to old growth often results from a rather inflexible requirement (e.g., large-diameter snags)
- species adapted to old growth usually show slower rates of reproduction than those adapted to younger successional stages, that is, they cannot recover quickly when their numbers are reduced.

There is insufficient information to characterize the degree of association with old-growth forests for most organisms – i.e., for most invertebrates, nonvascular plants, lichens, and macrofungi – in B.C. So little is known about microbes (including microfungi) that they are not treated in the following sections. A good introductory reference is Chanway (1993).

Birds

Species and Abundance

Lists of old growth-associated bird species are provided in Ruggiero et al. (1991) (species 'associated' or 'closely associated' with old growth); Bunnell and Chan-McLeod (1997) (species using old-growth habitat in coastal B.C.); FEMAT (1993), USDA Forest Service and USDI Bureau of Land Management (1994), and Thomas et al. (1993) (bird species closely associated with late successional forests within the range of the Northern Spotted Owl [*Strix occidentalis*]); Radcliffe et al. (1994) (bird species for which late seral stages in the Prince Rupert Forest Region [northwestern B.C.] have primary habitat value); and Fenger and Harcombe (1990) (B.C. bird species with 'suspected old growth forest habitat requirements'). Craig (in prep.) lists 122 bird species as breeding in old-growth forests in British Columbia.

A number of studies examined bird communities in younger and old-growth stands in the U.S. Pacific Northwest (e.g., Manuwal and Huff [1987], Huff et al. [1991], Lundquist and Mariani [1991], and Manuwal [1991] for Washington; Huff and Raley [1991] for Washington and Oregon; Carey et al. [1991] for Oregon; and Ralph et al. [1991] for Oregon and California). These studies found that, although old-growth stands often had a greater abundance of birds, there was little difference in species richness between stands of different ages. Manuwal (1991), for example, found that all but 3 of 17 bird species analyzed in Washington Douglas-fir forests were most abundant in old growth.

Many of these bird species were found to be associated with old growth but were also abundant in other seral stages (e.g., Western Flycatcher [*Empidonax difficilis*] and Varied Thrush [Carey et al. 1991]). Gilbert and

Allwine (1991a) suggest that most bird species do not differentiate between trees 200 years old and trees 450 years old. These studies for the Pacific Northwest Douglas-fir forests must be interpreted with caution, as 'young' and 'mature' stands were regenerated naturally following wildfire and so retained some attributes (snags, coarse woody debris, large live trees) of old-growth forests; in this way, they differ considerably from many managed second-growth forests. The same caution should be applied to similar results obtained by Savard et al. (in prep.) and Seip and Savard (in prep.) for coastal B.C. (Vancouver Island, Queen Charlotte Islands, and the south coast).

Studies of bird species abundance in different forest seral stages in British Columbia include Roe (1974), Buckner et al. (1975), Hatler et al. (1978), Bryant et al. (1993), Savard et al. (in prep.), and Seip and Savard (in prep.) for coastal B.C.; Seip (in review) and Seip (1996) for the central interior (Sub-Boreal Spruce Zone); Catt (1987) for Kootenay National Park; Wetmore et al. (1985) for subalpine forests on the coast (Mountain Hemlock Zone) and interior (Engelmann Spruce–Subalpine Fir Zone); Keisker (1987) and Harestad and Keisker (1989) for south-central B.C.; and Lance and Phinney (1993) and Farr (1994) for the boreal forest in northeastern B.C.

Schieck et al. (1995) found that two-thirds of bird species in aspen forests of northern Alberta had their highest abundances in older forests, and 11 bird species (Broad-winged Hawk [*Buteo platypterus*], Cooper's Hawk [*Accipiter cooperii*], Warbling Vireo [*Vireo gilvus*], Golden-crowned Kinglet [*Regulus satrapa*], Hairy Woodpecker [*Picoides villosus*], White-breasted Nuthatch [*Sitta carolinensis*], Brown Creeper [*Certhia americana*], Black-billed Cuckoo [*Coccyzus erythropthalmus*], Veery [*Catharus fuscescens*], Magnolia Warbler [*Dendroica magnolia*], and Blackpoll Warbler [*Dendroica striata*]) were found only in older forests. They relate this to greater spatial complexity in older aspen forests (e.g., Lee et al. 1995a, 1995b).

Cavity-Nesters
One obvious reason for old-growth association is a requirement, or preference, for large snags or live trees. These trees are used by primary cavity-nesters (species that excavate their own cavities), secondary cavity-nesters (species that use cavities they have not excavated themselves), and species that use the trees for other purposes, such as feeding, perching, and so on.

B.C.'s primary cavity-nesters are woodpeckers, sapsuckers, chickadees, and nuthatches. Larger living and dead trees are often used exclusively, or preferentially to smaller ones. The presence of decayed heartwood – surrounded by a harder, intact sapwood – is important for almost all primary cavity-nesters. Harestad and Keisker (1989), for example, found that the presence of heart-rot conks was a good predictor of tree use by cavity-nesters in south-central B.C. Erskine and McLaren (1972) discuss the use of larger-diameter

Cavity-nesters such as the Hairy Woodpecker
require larger living and dead trees that are most
commonly found in old growth. (Photograph by
Steve Voller)

aspen trees in early stages of heart-rot by Yellow-bellied Sapsuckers in
central-interior B.C.

Carey et al. (1991) report that cavity-nesters (including the Brown Creeper)
in the Oregon Coast Ranges use very large snags (mean diameter at breast
height [dbh] of 94 cm) for nesting. Schoen et al. (1988) note that 12 of the
26 old growth-associated bird species in southeastern Alaska rely on snags
for roosting and nesting, and that large snags (> 58 cm dbh) are used more
often. Habeck (1988) reports 8 bird species that require 'large, standing dead
trees' and 10 bird species that require 'large, living old-growth trees,' in the
Rocky Mountains. Bull et al. (1986) record Pileated (*Dryocopus pileatus*) and
Hairy woodpeckers, Northern Flickers (*Colaptes auratus*), and Williamson's
Sapsuckers (*Sphyrapicus thyroideus*) nesting in dead trees averaging more than
40 cm dbh and feeding in trees with mean dbh from 24 cm (Three-toed
Woodpecker [*Picoides tridactylus*]) to 34 cm (Black-backed Woodpecker

[*Picoides arcticus*]) and mean heights of 18-19 m. Williamson's Sapsucker nests in trees with mean dbh of 82 cm (Raphael and White 1984) and 70 cm (Bull et al. 1986). White-headed Woodpeckers (*Picoides albolarvatus*) require snags with a minimum dbh of 20 cm (Milne and Hejl 1989) and 25 cm (Bull 1978; Thomas 1979). Pileated Woodpeckers require trees of at least 25.8 cm dbh to nest (Campbell et al. 1990; Bull et al. 1992).

Westworth and Telfer (1993) note a 'pronounced increase' in cavity-nesting bird populations in aspen forests from 60- to 80-year-old stands, as they develop snags of sufficient size. In south-central B.C., Yellow-bellied Sapsuckers (*Sphyrapicus varius*) usually nested in trees more than 30 cm dbh, and Pileated Woodpeckers utilized trees more than 40 cm dbh for nesting (Keisker 1987). Williamson's Sapsucker feeds on trees with a mean dbh of 21 cm (Bull et al. 1986). Lundquist and Mariani (1991) found that numbers of Chestnut-backed Chickadees (*Parus rufescens*) and Vaux's Swifts (*Chaetura vauxi*) correlated strongly with density of live trees more than 100 cm dbh. Bald Eagles in Washington and Oregon roost communally at night in coniferous trees with dbh from 63 cm to 152 cm (Garrett et al. 1993; Isaacs et al. 1993), and nest in larger-diameter conifers close to the ocean on Barkley Sound and the Gulf Islands (Vermeer and Morgan 1989; Vermeer et al. 1989). Flammulated Owls (*Otus flammeolus*) (secondary cavity-nesters) appear to be restricted to old-growth Douglas-fir forests in B.C. (Howie and Ritcey 1987).

Primary cavity-nesters associated with old-growth forests include Black-backed Woodpeckers (Marshall 1992), Red-breasted Sapsuckers (*Sphyrapicus ruber*) (Carey et al. 1991; Manuwal 1991; Bryant et al. 1993; Savard et al., in prep.; Seip, in review), Hairy Woodpeckers (Zarnowitz and Manuwal 1985; Bryant et al. 1993; Savard et al., in prep.; Schieck et al., 1995), Pileated Woodpeckers (Campbell et al. 1990; Bull et al. 1992; Bryant et al. 1993; Seip, in review), Chestnut-backed Chickadees (Gilbert and Allwine 1991a; Huff et al. 1991; Lundquist and Mariani 1991; Manuwal 1991; Ralph et al. 1991; Bryant et al. 1993), and Red-breasted Nuthatches (*Sitta canadensis*) (Hatler et al. 1978; Huff et al. 1991; Lundquist and Mariani 1991; Ralph et al. 1991; Bryant et al. 1993; Savard et al., in prep.; Seip, in review). Backhouse and Lousier (1991) list all of B.C.'s sapsuckers and woodpeckers as highly dependent on wildlife trees for reproduction, feeding, and 'other factors.'

Secondary cavity-nesters associated with old-growth forests include Vaux's Swift (Gilbert and Allwine 1991a; Lundquist and Mariani 1991; Manuwal 1991; Bryant et al. 1993; Bull and Hohmann 1993; Savard et al., in prep.), Hammond's (*Empidonax hammondii*), Olive-sided (*Contopus borealis*), and Pacific-slope (*Empidonax difficilis*) (also was known as Western) flycatchers (Roe 1974; Carey et al. 1991; Gilbert and Allwine 1991a; Manuwal 1991; Ralph et al. 1991; Bryant et al. 1993; Savard et al., in prep.; Seip, in review), and most of B.C.'s owls.

Other Nesting Birds

Other bird species are associated with old-growth forests for reasons related to other attributes of large standing or dead trees. The Brown Creeper nests under loose bark on large, old trees (Carey 1989; Carey et al. 1991; Gilbert and Allwine 1991a; Huff and Raley 1991; Lundquist and Mariani 1991; Ralph et al. 1991; Bryant et al. 1993; Seip, in review; Schieck et al., 1995; Savard et al., in prep.); this is sometimes true for the Pacific-slope Flycatcher as well. Marbled Murrelets (*Brachyramphus marmoratus*) require large (> 80 cm dbh), broad-limbed nest trees (Sealy and Carter 1984; Rodway et al. 1992, 1993); these branches, with their thick moss coverings, may not develop until trees are more than 175 years old (USDA Forest Service and USDI Bureau of Land Management 1994).

Large logs are important forage sites for some bird species. Habeck (1988) reports that Northern Saw-whet (*Aegolius acadicus*) and Barred (*Strix varia*) Owls and the Pileated Woodpecker use coarse woody debris in this way. Pileated Woodpeckers use large-diameter (> 25 cm), long (> 15 m) Douglas-fir and western larch logs disproportionately to their occurrence (Bull et al. 1986).

Forest Type

Tree species can also be important: Lundquist and Mariani (1991) report (for the southern Cascade Range) that western white pine snags are important for woodpeckers and creepers, and Douglas-fir and western hemlock snags are important for chickadees and nuthatches. Craig (in prep.) reports that, in B.C., White-breasted and Pygmy (*Sitta pygmaea*) Nuthatches nest only in ponderosa pine, and the Western Bluebird (*Sialia mexicana*) nests almost entirely in ponderosa pine and Garry oak.

The use of old growth varies seasonally for some bird species. Manuwal and Huff (1987) and Huff et al. (1991) report that bird species richness, diversity, and abundance were all greater in old growth (compared with younger) Douglas-fir stands in winter but not in spring, when there were few differences. They suggest that old growth may provide better foraging and a more moderate microclimate in winter. Seed-eaters, such as Pine Siskins (*Carduelis pinus*) and crossbills, depend on conifer seeds for food over the winter. The increasing abundance of the later successional species western hemlock (which produces regular seed crops) in Douglas-fir (which produces irregular seed crops) stands may explain the association of siskins and crossbills with old-growth forests (Manuwal and Huff 1987; Carey 1989) in the Douglas-fir forests of Oregon and Washington. In coastal B.C., Red Crossbills (*Loxia curvirostra*) were found most abundantly (but not exclusively) in old growth (Roe 1974; Bryant et al. 1993; Savard et al., in prep.).

Finally, a few bird species may be associated with old-growth forest because this seral stage is often found in large blocks in areas lacking access roads. It has been suggested by a number of authors (e.g., Dunbar et al. 1991; Bart and Forsman 1992; FEMAT 1993; Thomas et al. 1993) that one of the requirements of the Northern Spotted Owl is unfragmented old forest – that fragmentation allows the intrusion of other species (especially the Barred Owl) that may outcompete the Northern Spotted Owl. Among bird species, Rosenberg and Raphael (1986) found that the Northern Spotted Owl and Pileated Woodpecker showed the greatest sensitivity to forest fragmentation; the Sharp-shinned Hawk (*Accipiter striatus*) and Blue Grouse (*Dendragapus obscurus*) were also of potential concern. USDA Forest Service and USDI Bureau of Land Management (1994) suggest that forest fragmentation may be responsible for the observed high failure rates of Marbled Murrelet nests. Increased visibility of nests and murrelets may lead to higher mortality among eggs, young, and adults by increasing predation from jays, crows, and ravens (USDA Forest Service and USDI Bureau of Land Management 1994).

Bryant et al. (1993) found that old-growth bird communities on Vancouver Island are composed largely of species that are year-round B.C. residents (25-41% of species winter outside B.C.), but communities in second-growth forest consist mainly of species that winter outside Canada (77-97% of species in 15- to 20-year-old forests winter outside B.C.). As Savard et al. (in prep.) note, since most of the bird studies are done during breeding season, the value of old growth versus second growth as winter habitat remains largely unknown.

Mammals

Lists of old growth-associated mammal species are provided in Ruggiero et al. (1991) (species 'associated' or 'closely associated' with old growth); Bunnell and Chan-McLeod (1997) (species using old-growth habitat in coastal B.C.); FEMAT (1993), USDA Forest Service and USDI Bureau of Land Management (1994), and Thomas et al. (1993) (mammalian species closely associated with late successional forests within the range of the Northern Spotted Owl); Radcliffe et al. (1994) (mammalian species for which late seral stages in the Prince Rupert Forest Region [northwestern B.C.] have primary habitat value); and Fenger and Harcombe (1990) (B.C. species with 'suspected old growth forest habitat requirements').

Mammals associated with old growth utilize the distinctive habitat components of old-growth forests, including large living and dead trees and coarse woody debris. In addition, some mammals require riparian areas and (for various reasons) forest interior habitat (for more information, see Chapters 4 and 5).

Insectivores

Among the insectivores, the shrew mole (*Neurotrichus gibbsii*) has been noted by a number of authors (Terry 1981; Raphael 1988; Corn and Bury 1991b) to be closely associated with old-growth forests. Other studies disagree (Corn et al. 1988; Gilbert and Allwine 1991b; West 1991), finding the species to be equally abundant in earlier seral stages. (Carey [1989] lists the shrew mole as 'degree of association uncertain.') This probably reflects the habitat requirements of the species (logs, thick soils rich in organic materials), which may be present in younger forests as well. Trowbridge's shrew (*Sorex trowbridgii*) is reported to be associated with old-growth forests (Gilbert and Allwine 1991b). Populations of vagrant shrews increase after logging, whereas populations of Trowbridge's shrews decline (Anthony and Morrison 1985; Corn et al. 1988; Raphael 1988). Shrew-mole and Trowbridge's shrew populations in the Lower Mainland of British Columbia are low and decreasing, respectively, primarily reflecting habitat loss and fragmentation (Zuleta and Galindo-Leal 1994). Another potential contributor to the decline of Trowbridge's shrew populations is the species' reliance on hypogeous fungi as a food item. Maser et al. (1978) found 100% of the fungal contents of Trowbridge's shrew digestive tracts to be hypogeous fungi in the Endogonaceae ('truffles'), which are abundant in older forests and generally absent from younger ones. Craig (in prep.) lists seven shrew species as breeding in old-growth forests in British Columbia. Roy et al. (1995) in northern Alberta and Seip and Savard (in prep.) in coastal B.C. found that masked shrews were associated with old-growth forests.

Rodents
Populations of deer mice (*Peromyscus maniculatus*), creeping voles (*Microtus oregoni*), and Townsend's chipmunks (*Tamias townsendii*) increase on a site after it is logged; red-backed voles decline (Sullivan and Krebs 1980; Anthony and Morrison 1985; Corn et al. 1988; West 1991). Almost all of these generalizations are contradicted by other studies. Research in the U.S. Pacific Northwest (Corn et al. 1988; Corn and Bury 1991b; Gilbert and Allwine 1991b) suggests that, although rodent abundance may peak in old-growth forests, few species occur exclusively there. This may be partly because (as noted for insectivores) these small mammals are likely responding to microenvironmental variables present in second-growth forests, especially in those regenerating naturally following wildfire. One exception to this may be the red-backed voles (*Clethrionomys* spp.) (Raphael 1988; Ralph et al. 1991; Roy et al. 1995). Northern and southern red-backed voles depend on hypogeous fungi for a considerable portion of their springtime diet (Maser et al. 1978; Ure and Maser 1982); these fungi are not present in any abundance in younger stands (Clarkson and Mills 1994). Seip (in review) notes that red-backed voles are most abundant in older forests in the central interior of B.C.

Carey (1989) speculates that there may be no old growth-dependent species among the small mammals dwelling on the forest floor, but recognizes northern flying squirrels (*Glaucomys sabrinus*) as 'closely associated with old growth' and Douglas' squirrel (*Taomiasciurus douglasii*) as 'degree of association uncertain.' Northern flying squirrels nest and den in large snags (Habeck 1988) and require a hardwood understorey (Raphael 1988). Both Witt (1992) and McDonald (1995) found greater densities of northern flying squirrels in old-growth (0.85 squirrels per hectare) than in second-growth (0.12 squirrels per hectare) forests; however, Rosenberg and Anthony (1992) found no differences in northern flying squirrel density, sex ratio, body mass, or annual recapture rate between old-growth and second-growth stands. (They suggest that northern flying squirrel populations in western Oregon may be kept in check through predation by Northern Spotted Owls.)

Douglas' squirrel reaches peak abundance in old-growth forests (Buchanan et al. 1990; Ralph et al. 1991), perhaps in response to greater and more reliable quantities of conifer seeds there. Densities of Townsend's chipmunk are greater in old growth than in second growth; this is related primarily to the density of large (> 50 cm dbh) snags (Rosenberg and Anthony 1992). Red squirrels are more abundant in older aspen stands (Roy et al. 1995), using snags and large trees for nesting and cover, and fungi and lichens growing on coarse woody debris as food. Fenger and Harcombe (1990) note that the mountain beaver needs mature and old-growth forest for reproduction and shelter. Craig (in prep.) lists 10 rodent species as breeding in old-growth forests in British Columbia.

Given the nature of their old-growth associations, and projected timber harvests, Raphael (1988) predicted long-term population declines in northwestern California for Douglas' squirrel (31%), northern flying squirrel (31%), dusky-footed woodrat (*Neotoma fuscipes*) (55%), and western red-backed vole (37%). Rosenberg and Raphael (1986) identified the northern flying squirrel as a species 'of potential concern' with respect to population decline resulting from forest fragmentation.

Carnivores

Lynx (*Lynx canadensis*) was on early lists compiled by the USDA Forest Service and USDI Bureau of Land Management (1994), but was dropped because 'it is not closely associated with old growth forests for much of its history.' Koehler and Aubry (1994) and Koehler and Brittell (1990) suggest that lynx need both early successional forests (as habitat for their prey, snowshoe hares [*Lepus americanus*]) and old-growth forests (which provide cover for denning and for concealing their kittens).

American martens (*Martes americana*) use old growth to avoid predators, to provide resting and denning places in coarse woody debris and large-diameter trees, to provide access under the snow surface, and to provide prey (Thompson and Harestad 1994; Thompson and Colgan 1994; Buskirk and Powell 1994; Koehler et al. 1990). American marten foraging success was 21-119% greater in old-growth versus successional forests (Thompson 1994). The use of second-growth forest by martens may be greater when coarse woody debris levels are high (Baker 1992). Given the nature of their old-growth association, and projected timber harvests, Raphael (1988) predicted a long-term decline of American marten populations in northwestern California of 26%.

Fishers (*Martes pennanti*) require old growth for the same reasons as martens (hunting, resting, and denning), utilizing large dead trees and coarse woody debris (Fenger and Harcombe 1990; Powell and Zielinski 1994). Rosenberg and Raphael (1986) identify fishers as being sensitive to forest fragmentation.

Wolverines are very secretive and require the coarse woody debris often found in old growth for denning and cover. (Courtesy B.C. Ministry of Environment, Lands and Parks)

Wolverines (*Gulo gulo*) shun people (Banci 1994), and in some areas require old-growth attributes (large windfalls for denning, coarse woody debris for prey habitat) (Hornocker and Hash 1981); in other studies the link with old growth was not evident (Gardner 1985; Banci 1987). River otters (*Lutra canadensis*) require large upturned trees for denning in some areas (Fenger and Harcombe 1990). Both black bears (*Ursus americanus*) and grizzly bears (*Ursus arctos*) utilize old-growth attributes such as large trees with root cavities, large snags, and coarse woody debris for feeding, resting, and denning, especially in areas of high snowfall (Fenger and Harcombe 1990). Craig (in prep.) lists 21 carnivore species as breeding in old-growth forests in British Columbia.

Ungulates
A large body of research, summarized in Nyberg and Janz (1990), documents deer and elk habitat in south coastal B.C. Deer here and elsewhere use older mature and old-growth forest in the winter (Schoen et al. 1984, 1988; Nyberg et al. 1986, 1987; Pauley et al. 1993; Stelfox et al. 1995) because of its reduced snow depths, moderate forage, and litterfall of arboreal lichens. This winter habitat requirement depends on the severity of the winter, and does not arise every year. For similar reasons, mountain goats (*Oreamnos americanus*) in wetter climates utilize old-growth forests (Fox et al. 1989; Fenger and Harcombe 1990), requiring forested areas adjacent to escape terrain. Old-growth forests are also critical to the winter survival of caribou (*Rangifer tarandus*) (Rominger and Oldemeyer 1989; Cichowski and Banner 1993), and to moose (*Alces alces*) in areas of deep snow (Fenger and Harcombe 1990).

Bats
Little is known about the ecology of bats in B.C. (Nagorsen and Brigham 1993). Most bat species in British Columbia require site attributes found most abundantly in old growth (large live trees, large snags, long snag longevity, loose bark, and crevices between bark plates), and most species are found in highest concentrations in old-growth forests (Perkins and Cross 1988; Thomas 1988; Thomas and West 1991; Christy and West 1993; Holroyd et al. 1994; Crampton and Barclay 1995). Bats appear to feed little in the old-growth forests; rather, the forests provide roosts (Thomas 1988; Thomas and West 1991). Big brown bats (*Eptesicus fuscus*) roost in large-diameter ponderosa pine snags in the southern interior of B.C. (Brigham 1991). Perkins and Cross (1988) suggest that, in Oregon, crevices in the bark of conifers that are used for roosts are not well developed before trees are 150 years old.

These roost trees are potentially important for the nitrogen nutrition of old-growth forests (Rainey et al. 1992). The USDA Forest Service and USDI

Bureau of Land Management (1994) recommend increasing late successional reserves, riparian protection, and retention of green trees, snags, and coarse woody debris to provide bat habitat. Craig (in prep.) lists 11 bat species as breeding in old-growth forests in British Columbia.

Reptiles

Lists of old growth-associated reptilian species are provided in FEMAT (1993), Thomas et al. (1993), and USDA Forest Service and USDI Bureau of Land Management (1994) (reptilian species closely associated with late successional forests within the range of the Northern Spotted Owl), and Blaustein et al. (1995). Fenger and Harcombe (1990), Ruggiero et al. (1991), and Thomas et al. (1993) found no reptilian species in the Pacific Northwest, including B.C., to be associated with old-growth forests. Craig (in prep.) lists four reptilian species as breeding in old-growth forests in British Columbia.

Few reptiles are associated with old-growth forests. In general, as forests are logged reptile populations increase and amphibian populations decrease (Raphael and Marcot 1986; Bury and Corn 1988a; Raphael 1988; Welsh and Lind 1991). Welsh and Lind (1991) report that sharp-tailed snakes (*Contia tenuis*) occur primarily in old-growth stands, but Raphael (1988) found this species primarily in younger stands. Sharp-tailed snakes are found in a variety of forested and non-forested habitats, and appear to feed mainly on slugs.

Amphibians

Lists of amphibian species associated with old growth are provided in Ruggiero et al. (1991) (species 'associated' or 'closely associated' with old growth); FEMAT (1993), Thomas et al. (1993), and USDA Forest Service and USDI Bureau of Land Management (1994) (old growth-associated amphibians within the range of the Northern Spotted Owl); Fenger and Harcombe (1990) (B.C. species with 'suspected old growth forest habitat requirements'); Radcliffe et al. (1994) (amphibian species for which late seral stages in the Prince Rupert Forest Region [northwestern B.C.] have primary habitat value); and Blaustein et al. (1995). Craig (in prep.) lists 10 species of newts and salamanders and 6 species of frogs as breeding in old-growth forests in British Columbia.

Because most amphibians are creatures of moist, cool habitats (at least for part of their life cycle), the conversion of old forests to younger, hotter, drier forests can have a major impact on local populations (Pough et al. 1987; Buhlman et al. 1988; Bury and Corn 1988b; Dupuis 1993). Amphibians are not particularly mobile creatures, and so dispersion to and from suitable habitats is limited. In addition, populations of many amphibian species are declining globally (Blaustein and Wake 1990) and locally (Welsh 1990; Walls et al. 1992). A good summary of Pacific Northwest amphibians and reptiles is provided in Blaustein et al. (1995).

A number of studies in the U.S. Pacific Northwest have shown that there are no significant differences among stand age classes with respect to amphibian species richness or diversity (Aubry and Hall 1991; Corn and Bury 1991a; Gilbert and Allwine 1991c; Welsh and Lind 1991), although a few (e.g., Welsh and Lind 1988) showed greater species richness in older forests.

Welsh and Lind (1991) showed that a number of salamander species were associated with microhabitat variables associated with late successional forests. Clouded salamanders (*Aneides ferreus*) are found in large coarse woody debris (decay classes 2 and 3) with loose bark, and their populations seem to be positively correlated with stand age (Stelmock and Harestad 1979; Aubry and Hall 1991; Corn and Bury 1991a; Gilbert and Allwine 1991c). Whitaker et al. (1986) point out that all common food items of the clouded salamander are found within rotting logs. Raphael (1988) estimates that, given habitat needs and currently scheduled cuts, clouded salamander populations will decline 29% in northwestern California.

Ensatinas (*Ensatina eschscholtzii*) are associated with more decayed coarse woody debris (Bury et al. 1991; Corn and Bury 1991a). They are not strongly associated with later successional stands (Aubry and Hall 1991; Corn and Bury 1991a; Thomas et al. 1993), although Welsh and Lind (1991) found them to be more abundant there. Similarly, western red-backed salamanders (*Plethodon vehiculum*) were found to be associated with old-growth forests by Corn and Bury (1991a) but not by most researchers (e.g., Aubry et al. 1988; Aubry and Hall 1991; Bury et al. 1991). Northwestern salamanders are associated with old growth in most studies (e.g., Aubry and Hall 1991; Bury et al. 1991; Seip and Savard, in prep.), but not in Corn and Bury (1991a). Rough-skinned newts (*Taricha granulosa*) seem to be associated with old growth (Aubry and Hall 1991; Corn and Bury 1991a; Thomas et al. 1993) or not (Corn and Bury 1991a; Gilbert and Allwine 1991c), depending on factors such as geography and site moisture. Much of the confusion here may result from the fact that these salamanders are all associated with coarse woody debris, and in a number of Pacific Northwest research sites, the amount of coarse woody debris was comparable in young and old stands.

Pacific giant salamanders (*Dicamptodon tenebrosus*) are not generally believed to be associated with old-growth forests (Bury and Corn 1988b), but they require clean, cool water for breeding. Rosenberg and Raphael (1986) rated the Pacific giant salamander as a species 'of potential concern' for decline resulting from forest fragmentation.

Tailed frogs (*Ascaphus truei*) also need cool, clean water for breeding, and are the amphibian species most affected by logging and habitat degradation (Blaustein et al. 1995). They are associated with old-growth forests (Raphael 1988; Welsh and Lind 1988; Aubry and Hall 1991; Seip and Savard, in prep.), but can persist in cutover areas if sufficient forested habitat remains upstream to keep the stream channel cool (Corn and Bury 1989). When logging alters stream temperatures, local extinction can result (Corn and Bury 1989). Western toads (*Bufo boreas*), Pacific treefrogs (*Hyla regilla*), and red-legged frogs (*Rana aurora*) are not restricted to old growth, but all seem to be most abundant there (Bury and Corn 1988a; Aubry and Hall 1991; Welsh and Lind 1991). This may result from the cooler, moister microclimate, but may also reflect the fact that these species are not subjected to pesticides or to the introduced American bullfrog there (*Rana catesbiano*) (Blaustein et al. 1995).

Fish
Damage to old-growth forests in riparian habitats is one factor contributing to the decline in west coast anadromous fish stocks (Nehlsen et al. 1991; USDA Forest Service and USDI Bureau of Land Management 1994). In this regard, salmon, steelhead (*Oncorhynchus mykiss*), and sea-run cutthroat (*Oncorhynchus clarki*) can be considered old growth-dependent fish. Because anadromous salmonids generally return to breed in their natal streams, they have a high degree of reproductive isolation from other 'stocks,' and are therefore considered genetically distinct. The management unit for conservation then becomes the stock rather than the species. The American Fisheries Council has produced a list of 214 native naturally spawning stocks from California, Oregon, Idaho, and Washington: 101 at high risk of extinction, 58 at moderate risk of extinction, 54 of special concern, and 1 classified as threatened under the Endangered Species Act of 1973 and as endangered by the state of California. Species on the list are chinook salmon (*Oncorhynchus tshawytscha*), coho salmon (*Oncorhynchus kisutch*), sockeye salmon (*Oncorhynchus nerka*), chum salmon (*Oncorhynchus keta*), pink salmon (*Oncorhynchus gorbuscha*), steelhead trout, and sea-run cutthroat trout. It is estimated that, over this area, 106 stocks have become extinct (FEMAT 1993).

The USDA Forest Service and USDI Bureau of Land Management (1994) list 7 fish races, species, or groups as being closely associated with late successional forests within the range of the Northern Spotted Owl.

No similar lists have been constructed for B.C., but the same risk factors identified for their range (degradation and loss of freshwater and estuarine habitats, timing and overexploitation in commercial and recreational fisheries, migratory impediments such as dams, and loss of genetic integrity reflecting the effects of hatchery practice and introduction of nonlocal stocks) apply in this province.

Invertebrates

The major terrestrial invertebrate phyla in B.C. are the Protozoa, Platy-helminthes, Rotifera, Nematoda, Mollusca, Annelida, Tardigrada, and Arthropoda.

Very little information is available regarding these groups in B.C. (Cannings 1994; Scudder 1994), but it is probable that they constitute the vast majority of species in the province. Relatively more information is available on the arthropods and molluscs, which are treated in more detail below.

Pojar (1993) estimates that B.C. contains about 600 species of spiders and 35,000 species of insects. Probably less than half of the invertebrate species in the province have been documented (Marshall et al. 1982; Cannings 1994); of these, many are likely to be new to science. Without a sound understanding of the provincial invertebrate fauna, it is impossible to estimate the number of species that rely on old-growth forests for habitat.

Arthropods

Lists of old growth-associated arthropod species are provided in Thomas et al. (1993) and in USDA Forest Service and USDI Bureau of Land Management (1994) (old growth-associated arthropods within the range of the Northern Spotted Owl). The latter reviews 15 functional groups of arthropods representing over 8,000 individual species. Arthropods (insects, crustaceans, arachnids, and myriapods) constitute over 85% of the biological diversity in late successional and old-growth forests in the Pacific Northwest (Asquith et al. 1990). Arthropods in old-growth forests are of interest for several reasons: (1) many of the species are flightless (and therefore have limited dispersal capabilities) (Lattin and Moldenke 1990; Moldenke and Lattin 1990a); (2) their flightless condition is believed to reflect habitat stability; (3) many of the old-growth forest associates have disjunct distributions and are found only in undisturbed forests (often endemic); (4) arthropods are a key to ecosystem function and may serve as indicators of ecosystem function; and (5) probably more than half of the species in our area are undescribed (USDA Forest Service and USDI Bureau of Land Management 1994).

Marshall (1993) estimates that there are at least 6,643 species of soil invertebrates in B.C. (including more than 2,000 species of mites and 3,000 species of insects), based on four different forest types (Coastal Douglas-Fir, Coastal Western Hemlock, Interior Douglas-Fir, and Engelmann Spruce–Subalpine Fir biogeoclimatic zones). Most soil arthropods show preferences for particular successional stages. Old-growth forests have the richest fauna, but no species were found to be restricted to old growth by Moldenke and Lattin (1990a, 1990b) and Setälä and Marshall (1994). However, Campbell and Winchester (1994) found that 8 species of staphylinid beetles are restricted to old-growth habitats in B.C. This suggests that distinct arthropod groups may behave differently.

Arthropod diversity in forest canopies has been documented by Schowalter (1989), Winchester (1997), and others. Canopy invertebrate communities in old-growth forests are rich in predatory and parasitic species, with complex food webs consisting of many trophic levels. Canopy invertebrate communities in second-growth forests are composed primarily of herbivorous and sucking species, with small populations of predators or parasites. Arthropods in old-growth forest canopies are poorly known; Winchester and Ring (1996) have discovered at least 65 species (5% of identified species) new to science by sampling five trees in the Carmanah Valley. This small sampling hints at the magnitude of the task in making an appropriate assessment of biodiversity.

As old-growth forests are converted to second-growth forests through logging and regeneration, populations of herbivorous arthropods in forest canopies will increase; this will have consequences for forest health. Forest health concerns involving arthropods may also result from fire suppression: the duration and intensity of western spruce budworm (*Choristoneura occidentalis*) outbreaks have increased with the decrease in forest fire frequency in western Montana since 1910, as Douglas-fir invades ponderosa pine sites (Anderson et al. 1987).

Molluscs
Molluscs in B.C. forests include land snails, slugs, aquatic snails, and clams. Many of these species are associated with old-growth forests or, probably more commonly, riparian areas in these forests.

Lists of old growth-associated mollusc species are provided in FEMAT (1993), Thomas et al. (1993), and USDA Forest Service and USDI Bureau of Land Management (1994) (old growth-associated molluscs within the range of the Northern Spotted Owl). The molluscan fauna of the Pacific Northwest is not well known; FEMAT (1993) estimate that more than half of the molluscan taxa within the range of the Northern Spotted Owl remain to be discovered. The 102 species of molluscs 'closely associated with late-successional forests' in the Pacific Northwest have a high degree of endemism as well (USDA Forest Service and USDI Bureau of Land Management 1994).

Vascular Plants
Lists of old growth-associated vascular plant species are provided in Ruggiero et al. (1991) (species 'associated' or 'closely associated' with old growth) and FEMAT (1993) (vascular plant species closely associated with late successional forests within the range of the Northern Spotted Owl). These lists contain only vascular plant species that occur in the U.S. Pacific Northwest.

Many vascular plants such as these orchids occur in greatest abundance in old-growth forests. As well, plants in old growth may differ chemically from those found in younger forests. (Photograph by Joan Voller)

Some of the unique or most characteristic features of old-growth forests involve vascular plants, including large trees, living and dead, horizontal and vertical. Other chapters in this book cover dead trees, standing (snags) and on the ground (coarse woody debris). Most of this chapter treats old-growth forests as habitat for other organisms; this section, however, characterizes such forests as habitat for the vascular plant species that make up an old-growth habitat.

While most vascular plants are annuals or biennials, most such plants in B.C.'s old-growth forests are perennials. The longest-lived of the perennials are trees, which can reach ages of 1,000 years or more (Pojar and MacKinnon 1994). The structural attributes of old-growth forests are dominated by these large, old trees: in 175-250 years on the coast (Franklin et al. 1981), old-growth forests develop their characteristic attributes of large, living old trees; large snags; large logs on land (coarse woody debris); and large logs in streams.

The structural attributes of old-growth forest types similar to those found in B.C. are discussed in Franklin and Waring (1980), Franklin et al. (1981), Spies and Cline (1988), Franklin and Spies (1991), and Spies and Franklin (1991) for Douglas-fir–western hemlock forests in the U.S. Pacific Northwest; Alaback (1984a) for western hemlock–Sitka spruce old growth in southeastern Alaska; Habeck (1978), Turner and Franz (1985), and Moeur (1992) for western redcedar–western hemlock forests in northern Idaho and Washington; Habeck (1990) and Arno et al. (1995) for ponderosa pine-dominated stands in Montana; Johnson and Fryer (1989) for subalpine lodgepole pine–Engelmann spruce forests in southwestern Alberta; Lee et al. (1995a, 1995b) for aspen mixed-wood stands in northern Alberta; Kneeshaw (1992) and Clark (1992) for hybrid white spruce–subalpine fir forests in the central interior of B.C.; and Arsenault and Bradfield (1995) and Gagnon and Bradfield (1986, 1987) for different forest types on the B.C. coast.

Vascular plant succession leading to an old-growth stage has been described by Halpern (1988), Schoonmaker and McKee (1988), Spies (1991a), Spies and Franklin (1991), and Halpern and Spies (1995) for the U.S. Pacific Northwest; Alaback (1982, 1984b) for southeastern Alaska; Habeck (1968) for northern Montana; Lee et al. (1995c) for aspen mixed-wood stands in northern Alberta; Clark (1992) for hybrid white spruce–subalpine fir forests in the central interior of B.C.; and Lertzman (1992) and Lertzman and Krebs (1991) for subalpine forests in southwestern B.C. In general, forests go through an early successional forb/shrub stage, where vascular plant species diversity is high but quite different from that of old growth: Schoonmaker and McKee (1988) found that 'late seral species' accounted for 99% of the cover in old-growth stands, 40% 5 years after cutting, and back up to 97% after 40 years. This vascular plant community stability is discussed by Halpern (1988). Vascular plant species diversity drops rapidly following canopy closure and understorey exclusion in some ecosystems (Alaback 1982, 1984b).

Although the USDA Forest Service and USDI Bureau of Land Management (1994) estimate that 124 vascular plant species in the Pacific Northwest are 'closely associated with late-successional forests,' few such species appear to be associated exclusively with old-growth forests. Franklin et al. (1981) and B.C. Ministry of Forests (1992) suggest that saprophytic and mycorrhizal vascular plants may require old-growth forests as habitat, reflecting the presence of large, old trees and humus-rich soils; Schoonmaker and McKee (1988) report that western coralroot (*Corallorhiza maculata*) and pinesap (*Hypopitys monotropa*) appeared to be eradicated following disturbance. Ruggiero et al. (1991) report 3 species of vascular plants (western yew [*Taxus brevifolia*], vanilla leaf [*Achlys triphylla*], and bunchberry [*Cornus canadensis*]) as 'closely associated' with old growth, and an additional 18 species as 'associated,' in Douglas-fir forests of the U.S. Pacific Northwest.

A number of vascular plant species occur in greatest abundance in old-growth forests, however. Spies (1991a, 1991b) and Ruggiero et al. (1991) suggest that western yew and *Tiarella* spp. are strongly associated with old growth. Halpern and Spies (1995) identify a number of vascular plant species (including western yew) as significantly associated with old-growth Douglas-fir forests. Klinka et al. (1985) suggest that a number of vascular (and nonvascular) plant species can be used to characterize old-growth forests in the Dry Maritime Coastal Western Hemlock (CWHdm) subzone in B.C.: heart-leaved twayblade (*Listera cordata*), false-azalea (*Menziesia ferruginea*), clasping-leaved twistedstalk (*Stretopus amplexifolius*), Alaska (*Vaccinium alaskaense*) and oval-leaved (*Vaccinium ovalifolium*) blueberries. They note that, as stands age, thicker, moister, and more acidic forest floors develop, leading to the presence of these species.

Vascular plants in old-growth forests may differ chemically from those in younger forests. Hanley et al. (1987), for example, report that plants of Alaska blueberry and bunchberry from young stands had greater astringency and higher concentrations of phenolics and non-structural carbohydrates, but lower concentrations of nitrogen, than plants from older stands.

Nonvascular Plants

Lists of old growth-associated nonvascular plant species are provided in Ruggiero et al. (1991) (species 'associated' or 'closely associated' with old growth); FEMAT (1993) and USDA Forest Service and USDI Bureau of Land Management (1994) (nonvascular plant species closely associated with late successional forests within the range of the Northern Spotted Owl); Lesica et al. (1991); and Soderstrom (1989).

Nonvascular plants include bryophytes and algae, but so little is known about B.C.'s terrestrial and freshwater algae that they are not discussed further here.

Bryophytes are important components of most forest ecosystems in B.C., contributing substantially to biomass production and nutrient cycling within these ecosystems (Pike et al. 1972; Binkley and Graham 1981). Old-growth forests often have distinctive bryophyte communities, resulting largely from microclimate modification and provision of habitat associated with late successional forests.

Lesica et al. (1991) cite 8 species of leafy liverworts (7 of which occurred only in old growth in their area) as possible old-growth associates in an old-growth grand fir stand in Montana. They suggest that higher humidity resulting from a deeper canopy and greater amounts of water-holding woody debris is required for some of these liverworts. Soderstrom (1988) reports similar results from old-growth spruce stands in Sweden: of epixylic liverworts, 6 species occur in old-growth stands and none occur in younger managed stands. McCune and Antos (1982) agree that moisture seems to be the key factor for most nonvascular epiphytes. Crites and Dale (1995) found sensitive liverwort species only on aspen logs in intermediate stages of decay.

Mosses seem not so strongly tied to old-growth habitats, although a number of species occur there in greatest abundance. McCune and Antos (1981b) cite 9 moss species as possible old-growth associates. Klinka et al. (1985) use *Rhizomnium glabrescens* and *Rhytidiadelphus loreus* to characterize old-growth forests in the CWHdm subzone in southwestern B.C., noting that these species showed up only when the forest floor developed greater moisture-holding capacity, incorporated more wood, and became more acidic. Clark (1992) reported that total moss cover increases with stand age in hybrid white spruce–subalpine fir forests in the central interior of B.C.

Substrate is also important. Soderstrom (1988, 1989) found 4 epixylic bryophyte species that occurred only on well-decayed wood; 6 species were found only on logs with little or no bark, and 7 species were restricted to logs more than 20 cm in diameter. Crites and Dale (1995), working in aspen stands in northern Alberta, found that different decay classes of logs had different abundances of bryophyte species: epiphytes were more abundant on early decay classes, epixylics on intermediate decay classes, and terricolous species on more decayed coarse woody debris.

And as noted by Soderstrom (1989), Lesica et al. (1991), Gustafsson et al. (1992a), Berg et al. (1994), Selva (1994), and Crites and Dale (1995), some bryophyte species have very limited dispersal capabilities, and in a highly fragmented landscape may be unable to recolonize stands following disturbance. For these species, the important variable may simply be time between disturbances (also referred to, in a temporal sense, as 'environmental continuity').

Fungi
Lists of old growth-associated fungal species are provided in Thomas et al. (1993) and USDA Forest Service and USDI Bureau of Land Management (1994) (fungal species closely associated with late successional forests within the range of the Northern Spotted Owl), and in Ruggiero et al. (1991).

Fungal biodiversity is poorly known for B.C. (Redhead 1994; Redhead and Berch 1997). Redhead (1994) reports over 1,250 species (likely a small fraction of the total) of macrofungi documented for the province. Because old-growth forests contain many attributes known to be important for fungi (e.g., dead and dying trees and logs, a variety of hosts of different species and ages, tree roots), these forests are believed to be an important fungal habitat. The USDA Forest Service and USDI Bureau of Land Management (1994) list 527 fungal species as being 'closely associated with late-successional forests.' Redhead and Berch (1997) recommend establishing

The fungi *Phaeocollybia attenuata* (top) and *Gomphus kauffmanii* (bottom) are west coast endemic species found mostly in old growth. (Photographs by Scott Redhead)

permanent fungal sample plots in representative old-growth habitats throughout B.C.

Hypogeous (truffle) fungi are particularly abundant in Pacific Northwest coniferous forests, where they form ectomycorrhizae with the roots of vascular plants (Perry et al. 1987; Molina et al. 1992). Ruggiero et al. (1991) report that six species of hypogeous fungi are associated with old-growth Douglas-fir forests in the U.S. Pacific Northwest. Species diversity and biomass production in hypogeous fungi appear not to differ with age in coastal Douglas-fir stands. Luoma (1991) reports that 1.3 kg/ha of sporocarps are produced annually, with no difference between species in younger and older stands. Luoma and Frenkel (1991) note a species shift, with young mesic stands dominated by *Gautieria monticola* and older mesic stands not. On the other hand, Clarkson and Mills (1994) found truffles much more abundant in old-growth forests, especially near logs.

Two species of the mushroom genus *Phaeocollybia* are known in Canada only from old-growth hemlock forests in the Carmanah Valley on Vancouver Island's west coast (Redhead and Norvell 1993). Although the genus is probably more widely distributed in B.C., there is reason to believe that it may be restricted to old-growth habitats.

The preceding paragraphs deal with macrofungi. B.C.'s microfungi are largely unknown (Pojar 1993), and no realistic estimates of their numbers, let alone their old-growth dependency, can be made.

Lichens

Lists of old growth-associated lichen species are provided in Goward (1993a, 1993b, 1994a, 1994b), Goward et al. (1994), Goward and McCune (in prep.), Ruggiero et al. 1991 (species 'associated' or 'closely associated' with old growth); Thomas et al. (1993); and FEMAT (1993) (lichen species closely associated with late successional forests within the range of the Northern Spotted Owl).

Lichens perform a variety of roles in B.C.'s forests: they are primary producers; cyanolichens fix and 'leak' nitrogen; they retain moisture within forest canopies; and they provide an important seasonal food source for a variety of animals, from small mammals to ungulates. In addition, many of our species are endemic to western North America, and a number of these appear to be restricted in their distribution to old-growth forests.

Goward (1993a) identifies 44 species restricted in the Coastal Western Hemlock (CWH) Zone to old-growth forests, and notes that age-related floristic differences are greater here than in most of the province. Forests may take 175-200 years to slow their growth and establish a multi-storeyed canopy, prerequisites for some of these lichens; it may take several centuries after that for them to acquire their complement of lichen species. Goward

(1993b, 1994b) also reports 24 species of macrolichens from interior wetbelt (Interior Cedar-Hemlock [ICH] Zone) old-growth forests that were hitherto known only from coastal B.C. Diversity appears to be positively correlated with forest age, and older old-growth forests (> 300-350 years), which Goward (1993a) calls 'antique' forests, have greater lichen species diversity than younger old-growth forests. Lichen diversity here is correlated with environmental continuity: since successful long-distance dispersal of these lichens occurs very infrequently, older old-growth forests have more species than younger forests. These generally lower-canopy, hygrophytic cyanolichens may be relicts from the 'Little Ice Age,' when climatic conditions would have been more favourable for their long-distance transport. In both the ICH and CWH zones, the important factor appears to be the age of the forest, not the age of the individual trees (Gustafsson et al. 1992a, 1992b; Berg et al. 1994; Goward 1994b).

Lesica et al. (1991) found that old-growth grand fir forests in Montana provide optimum habitat for many species of lichens, especially the nitrogen-fixing foliose lichens, *Alectoria* spp., and *Lobaria pulmonaria* (which was found only in old growth). Primary factors appear to be large amounts of coarse woody debris and a multilayered canopy (both of which provide humid conditions), and long-term continuity of woody vegetation. Pike

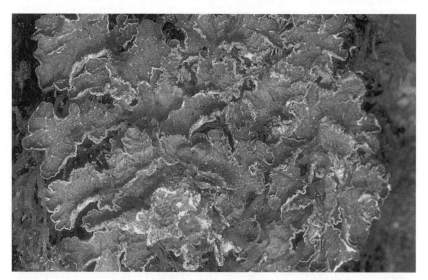

Lichens perform a variety of roles in British Columbia's forests. They are primary producers, they retain moisture within forest canopies, and they provide an important seasonal food source for a variety of animals. (Photograph by Anna Roberts)

et al. (1977) measured almost 18 kg of epiphytes on an old-growth Douglas-fir tree in Oregon; over 10 kg consisted of *Lobaria oregana.* Halpern and Spies (1995) also report that *Lobaria* species are associated with old-growth Douglas-fir forests in the Pacific Northwest.

McCune and Antos (1981a, 1981b) found that open canopies in drier climates were probably optimal for 'leafy' macrolichens, allowing wetting and drying. Foliose macrolichens decreased under closed canopies in their area, but several crustose lichens (especially in the Caliciales) increased.

Research Needs

Research needs with respect to biodiversity and old-growth forests can be divided into three categories:

What elements of biodiversity are associated with old-growth forests in B.C.? Basic inventories are still required to document the species (especially in lesser-known groups such as microbes, invertebrates, fungi, and algae) that appear to require old-growth forests as habitat.

What is the nature of dependency on old growth? On what feature of old-growth forests – large live or dead standing trees, coarse woody debris, unroaded areas, time, and so on – do species found solely or predominantly in such forests depend? Why?

How can forest management accommodate these dependencies in order to maintain acceptable levels of biodiversity in the province? Important questions include:

- Can we manage forests in such a way that habitat is provided for the various species that depend on old-growth attributes?
- How do we manage forests when we are hampered by an inadequate knowledge of the diversity present in them and the interdependence of the organisms?
- How do we learn from our experiences in a systematic fashion?

Another aspect requiring further research is the winter use of old-growth forests by birds: because most bird studies occur during the breeding season, the value of old growth compared with second growth as winter habitat is largely unknown.

Literature Cited

Alaback, P.B. 1982. Dynamics of understory biomass in Sitka spruce–western hemlock forests of southeast Alaska. Ecology 63(6):1932-1948.

–. 1984a. A comparison of old-growth and second-growth forest structure in the western hemlock–Sitka spruce forests of southeastern Alaska. Pages 219-226 in W.R. Meehan, T.R. Merrell Jr., and T.A. Handley (eds.). Fish and wildlife relationships in old-growth forests. Proceedings of a symposium. 12-15 April 1982, Juneau, Alaska. American Institute of Fishery Research Biologists, Moorehead City, N.C. 425 pp.

–. 1984b. Plant succession following logging in the Sitka spruce–western hemlock forests of southeast Alaska: implications for management. USDA Forest Service General Technical Report PNW-173. Portland, Ore.

Anderson, L., C.E. Carlson, and R.H. Wakimoto. 1987. Forest fire frequency and western spruce budworm outbreaks in western Montana. Forest Ecology and Management 22:251-260.

Anthony, R.G., and M.L. Morrison. 1985. Influence of glyphosate herbicide on small mammal populations in western Oregon. Northwest Science 59:159-168.

Arno, S.F., J.H. Scott, and M.G. Hartwell. 1995. Age class structure of old growth ponderosa pine/Douglas-fir stands and its relationship to fire history. USDA Forest Service Research Paper INT-RP-481. Ogden, Utah.

Arsenault, A., and G. Bradfield. 1995. Structural-compositional variation in three age-classes of temperate rainforests in southern coastal British Columbia. Canadian Journal of Botany 73(1):54-64.

Asquith, A., J.D. Lattin, and A.R. Moldenke. 1990. Arthropods: the invisible diversity. Northwest Environmental Journal 6(2):404-405.

Aubry, K.B., and P.A. Hall. 1991. Terrestrial amphibian communities in the southern Washington Cascade Range. Pages 327-338 in L.F. Ruggiero, K.B. Aubry, A.B. Carey, and M.H. Huff (eds.). Wildlife and vegetation of unmanaged Douglas-fir forests. USDA Forest Service General Technical Report PNW-GTR-285. Portland, Ore.

Aubry, K.B., L.L.C. Jones, and P.A. Hall. 1988. Use of woody debris by plethodontid salamanders in Douglas-fir forests in Washington. Pages 32-37 in R.C. Szaro, K.E. Severson, and D.R. Patton (tech. coords.). Management of amphibians, reptiles and small mammals in North America. USDA Forest Service General Technical Report RM-166. Fort Collins, Colo.

Backhouse, F., and J.D. Lousier. 1991. Silviculture systems research: wildlife tree problem analysis. B.C. Ministry of Forests, B.C. Ministry of Environment, and B.C. Wildlife Tree Committee, Victoria, B.C. 205 pp.

Baker, J.M. 1992. Habitat use and spatial organization of pine marten on southern Vancouver Island, British Columbia. M.Sc. thesis, Simon Fraser University, Burnaby, B.C.

Banci, V. 1987. Ecology and behaviour of wolverine in the Yukon. M.Sc. thesis, Simon Fraser University, Burnaby, B.C.

Banci, V. 1994. Wolverine. Pages 99-127 in L.F. Ruggiero, K.B. Aubry, S.W. Buskirk, L.J. Lyon, and W.J. Zielinski (tech. eds.). The scientific basis for conserving forest carnivores. American marten, fisher, lynx, and wolverine in the western United States. USDA Forest Service General Technical Report RM-254. Fort Collins, Colo. 184 pp.

Bart, J., and E.D. Forsman. 1992. Dependence of Northern Spotted Owls *Strix occidentalis caurina* on old-growth forests in the western USA. Biological Conservation 62:95-100.

Berg, Å., B. Ehnström, L. Gustafsson, T. Hallingbäck, M. Jonsell, and J. Weslien. 1994. Threatened plant, animal and fungal species in Swedish forests: distribution and habitat associations. Conservation Biology 8(3):718-731.

Binkley, D., and R.L. Graham. 1981. Biomass, production, and nutrient cycling of mosses in an old-growth Douglas-fir forest. Ecology 62(5):1387-1389.

Blaustein, A.R., and D.B. Wake. 1990. Declining amphibian populations: a global phenomenon? Trends in Ecology and Evolution 5(7):203-204.

Blaustein, A.R., J.J. Beatty, D.H. Olson, and R.M. Storm. 1995. The biology of amphibians and reptiles in old-growth forests in the Pacific Northwest. USDA Forest Service General Technical Report PNW-GTR-337. Portland, Ore.

Brigham, M.R. 1991. Flexibility in foraging and roosting behaviour by the Big Brown Bat (*Eptesicus fuscus*). Canadian Journal of Zoology 69:117-121.

B.C. Ministry of Forests. 1992. An old growth strategy for British Columbia. Integrated Resource Policy Branch, Victoria, B.C.

Bryant, A.A., J.-P.L. Savard, and R.T. McLaughlin. 1993. Avian communities in old-growth and managed forests of western Vancouver Island, British Columbia. Technical Report Series No. 167. Canadian Wildlife Service, Pacific and Yukon Region, British Columbia.

Buchanan, J.B., R.W. Lundquist, and K.B. Aubry. 1990. Winter populations of Douglas' squirrels in different-aged Douglas-fir forests. Journal of Wildlife Management 54(4):577-581.

Buckner, C.H., A.J. Erskine, R. Lidstone, B.B. McLeod, and M. Ward. 1975. The breeding bird communities of coast forest stands on northern Vancouver Island. Murrelet 56(3):6-11.

Buhlman, K.A., C.A. Pague, J.C. Mitchell, and R.B. Glasgow. 1988. Forestry operations and terrestrial salamanders: techniques in a study of the Cow Knob salamander, *Plethodon punctatus*. Pages 38-44 in R.C. Szaro, K.E. Severson, and D.R. Patton (tech. coords.). Management of amphibians, reptiles and small mammals in North America. USDA Forest Service General Technical Report RM-166. Fort Collins, Colo.

Bull, E.L. 1978. Specialized habitat requirements of birds: snag management, old growth, and riparian habitat. Pages 74-82 in R.M. DeGraaf (tech. coord.). Nongame bird habitat management in the coniferous forests of the western United States: proceedings of a workshop. 7-9 February 1977. USDA Forest Service General Technical Report PNW-GTR-64. Portland, Ore.

Bull, E.L., and J.E. Hohmann. 1993. The association between Vaux's Swifts and old-growth forests in northeastern Oregon. Western Birds 24:38-42.

Bull, E.L., R.S. Holthausen, and M.G. Henjum. 1992. Roost trees used by Pileated Woodpeckers in northeastern Oregon. Journal of Wildlife Management 56(4):786-793.

Bull, E.L., S.R. Peterson, and J.W. Thomas. 1986. Resource partitioning among woodpeckers in northeastern Oregon. USDA Forest Service Research Note PNW-444. LaGrande, Ore.

Bunnell, F.L., and A.C. Chan-McLeod. 1997. Terrestrial vertebrates. Pages 103-130 in Schoonmaker, P., B. von Hagen, and E.C. Wolf (eds.). The rain forests of home: profile of a North American bioregion. Island Press, Washington, D.C.

Bunnell, F.L., and L.L. Kremsater. 1990. Sustaining wildlife in managed forests. Northwest Environmental Journal 6:243-269.

Bury, R.B., and P.S. Corn. 1988a. Douglas-fir forests in the Oregon and Washington Cascades: relation of the herpetofauna to stand age and moisture. Pages 11-22 in R.C. Szaro, K.E. Severson, and D.R. Patton (tech. coords.). Management of amphibians, reptiles and small mammals in North America. USDA Forest Service General Technical Report RM-166. Fort Collins, Colo.

Bury, R.B., and P.S. Corn. 1988b. Responses of aquatic and streamside amphibians to timber harvest: a review. Pages 165-181 in K.J. Raedeke (ed.). Streamside management: riparian wildlife and forestry interactions. Contribution No. 59, Institute of Forest Resources, University of Washington, Seattle, Wash.

Bury, R.B., P.S. Corn, and K.B. Aubry. 1991. Regional patterns of terrestrial amphibian communities in Oregon and Washington. Pages 341-350 in L.F. Ruggiero, K.B. Aubry, A.B. Carey, and M.H. Huff (eds.). Wildlife and vegetation of unmanaged Douglas-fir forests. USDA Forest Service General Technical Report PNW-GTR-285. Portland, Ore.

Buskirk, S.W., and R.A. Powell. 1994. Habitat ecology of fishers and American martens. Pages 283-296 in S.W. Buskirk, A.S. Harestad, M.G. Raphael, and R.A. Powell (comps. and eds.). Martens, sables, and fishers: biology and conservation. Cornell University Press, Ithaca, N.Y. 484 pp.

Campbell, J.M., and N.N. Winchester. 1994. First record of *Pseudohaida rothi* Hatch (Coleoptera:Staphylinidae:Omaliinae) from Canada. Journal of the Entomological Society of British Columbia 90:83.

Campbell, R.W., N.K. Dawe, I. McTaggart-Cowan, J.M. Cooper, G.W. Kaiser, and M.C.E. McNall. 1990. The birds of British Columbia. 2 volumes. Royal British Columbia Museum, Victoria, B.C.

Cannings, S. 1994. Endangered terrestrial and freshwater invertebrates in British Columbia. Pages 47-52 in L.E. Harding, and E. McCullum (eds.). Biodiversity in British Columbia. Our changing environment. Environment Canada/Canadian Wildlife Service, Ottawa, Ont.

Carey, A.B. 1984. A critical look at the issue of species-habitat dependency. Pages 346-351 in New forests for a changing world: Proceedings of the annual convention of the Society of American Foresters, Portland, Oregon. Society of American Foresters, Bethesda, Md.

–. 1989. Wildlife associated with old-growth forests in the Pacific Northwest. Natural Areas Journal 9(3):151-162.

Carey, A.B., M.M. Hardt, S.P. Horton, and B.L. Biswell. 1991. Spring bird communities in the Oregon Coast Range. Pages 123-142 in L.F. Ruggiero, K.B. Aubry, A.B. Carey, and M.H. Huff (eds.). Wildlife and vegetation of unmanaged Douglas-fir forests. USDA Forest Service General Technical Report PNW-GTR-285. Portland, Ore.

Catt, D.J. 1987. Bird communities and forest succession in the subalpine zone of Kootenay National Park, British Columbia. M.Sc. thesis, Simon Fraser University, Burnaby, B.C.

Chanway, C.P. 1993. Biodiversity at risk: soil microflora. Pages 229-238 in M.A. Fenger, E.H. Miller, J.F. Johnson, and E.J.R. Williams (eds.). Our living legacy. Proceedings of a symposium on biological diversity. Royal British Columbia Museum, Victoria, B.C.

Christy, R.E., and S.D. West. 1993. Biology of bats in Douglas-fir forests. USDA Forest Service General Technical Report PNW-GTR-308. Corvallis, Ore.

Cichowski, D.B., and A. Banner. 1993. Management strategy and options for the Tweedsmuir-Entiako caribou winter range. Land Management Report Number 83. B.C. Ministry of Forests, Research Branch, Victoria, B.C.

Clark, D.F. 1992. Post-fire succession in the sub-boreal spruce forests of the Nechako Plateau, central British Columbia. M.Sc. thesis, University of Victoria, Victoria, B.C.

Clarkson, D.A., and L.S. Mills. 1994. Hypogeous sporocarps in forest remnants and clearcuts in southwest Oregon. Northwest Science 68(4):259-265.

Corn, P.S., and R.B. Bury. 1989. Logging in western Oregon: responses of headwater habitats and stream amphibians. Forest Ecology and Management 29:39-57.

–. 1991a. Terrestrial amphibian communities in the Oregon Coast Range. Pages 305-317 in L.F. Ruggiero, K.B. Aubry, A.B. Carey, and M.H. Huff (eds.). Wildlife and vegetation of unmanaged Douglas-fir forests. USDA Forest Service General Technical Report PNW-GTR-285. Portland, Ore.

–. 1991b. Small mammal communities in the Oregon Coast Range. Pages 241-254 in L.F. Ruggiero, K.B. Aubry, A.B. Carey, and M.H. Huff (eds.). Wildlife and vegetation of unmanaged Douglas-fir forests. USDA Forest Service General Technical Report PNW-GTR-285. Portland, Ore.

Corn, P.S., R.B. Bury, and T.A. Spies. 1988. Douglas-fir forests in the Cascade Mountains of Oregon and Washington: is the abundance of small mammals related to stand age and moisture? Pages 340-352 in R.C. Szaro, K.E. Severson, and D.R. Patton (tech. coords.). Management of amphibians, reptiles and small mammals in North America. USDA Forest Service General Technical Report RM-166. Fort Collins, Colo.

Craig, V. (in prep.). Breeding habitat of the forest-dwelling vertebrates of British Columbia. B.C. Ministry of Forests, Victoria, B.C.

Crampton, L.H., and R.M.R. Barclay. 1995. Relationships between bats and stand age and structure in aspen mixedwood forests in Alberta. Pages 211-225 in J.B. Stelfox (ed.). Relationships between stand age, stand structure, and biodiversity in aspen mixedwood forests in Alberta. Jointly published by Alberta Environmental Centre (AECV95-R1), Vegreville, Alta., and Canadian Forest Service (Project No. 0001A), Edmonton, Alta. 308 pp.

Crites, S., and M. Dale. 1995. Relationships between nonvascular species and stand age and structure in aspen mixedwood forests in Alberta. Pages 91-114 in J.B. Stelfox (ed.). Relationships between stand age, stand structure, and biodiversity in mixedwood forests in Alberta. Jointly published by Alberta Environmental Centre (AECV95-R1), Vegreville, Alta., and Canadian Forest Service (Project No. 0001A), Edmonton, Alta. 308 pp.

Dunbar, D.L., B.P. Booth, E.D. Forsman, A.E. Hetherington, and D.J. Wilson. 1991. Status of the Spotted Owl, *Strix occidentalis*, and Barred Owl, *Strix varia*, in southwestern British Columbia. Canadian Field Naturalist 105(4):464-468.

Dupuis, L.A. 1993. The status and distribution of terrestrial amphibians in old-growth forests and managed stands. M.Sc. thesis, University of British Columbia, Vancouver, B.C.

Erskine, A.J., and W.D. McLaren. 1972. Sapsucker nest holes and their use by other species. Canadian Field Naturalist 86(4):357-361.

Farr, D. 1994. Monitoring changes in wildlife diversity during operational hardwood harvesting – aspen clearcutting in the Dawson Creek Forest District. 1993/94 Draft Progress Report. B.C. Ministry of Environment, Lands and Parks, Wildlife Branch, Victoria, B.C.

Fenger, M., and A. Harcombe. 1990. Biodiversity, old growth forests and wildlife in British Columbia. Unpublished report, B.C. Ministry of Environment, Lands and Parks, Wildlife Branch, Victoria, B.C.

Forest Ecosystem Management Assessment Team (FEMAT). 1993. Forest ecosystem management: an ecological, economic and social assessment. USDA Forest Service (and other agencies), Portland, Ore.

Fox, J.L., C.A. Smith, and J.W. Schoen. 1989. Relation between mountain goats and their habitat in southeastern Alaska. USDA Forest Service General Technical Report GTR-246. Portland, Ore.

Franklin, J.F., and T.A. Spies. 1991. Composition, function, and structure of old-growth Douglas-fir forests. Pages 71-80 in L.F. Ruggiero, K.B. Aubry, A.B. Carey, and M.H. Huff (eds.). Wildlife and vegetation of unmanaged Douglas-fir forests. USDA Forest Service General Technical Report PNW-GTR-285. Portland, Ore.

Franklin, J.F., and R.H. Waring. 1980. Distinctive features of the northwestern coniferous forest: development, structure, and function. Pages 59-86 in R.H. Waring (ed.). Forests: fresh perspectives from ecosystem analysis. Proceedings of the 40th Annual Biology Colloquium. Oregon State University Press, Corvallis, Ore.

Franklin, J.F., K. Cromack Jr., W. Denison, A. McKee, C. Maser, J. Sedell, F. Swanson, and G. Juday. 1981. Ecological characteristics of old-growth Douglas-fir forests. USDA Forest Service General Technical Report PNW-118. Portland, Ore.

Gagnon, D., and G.E. Bradfield. 1986. Relationships among forest strata and environment in southern coastal British Columbia. Canadian Journal of Forest Research 16:1264-1271.

–. 1987. Gradient analysis of west central Vancouver Island forests. Canadian Journal of Botany 65:822-833.

Gardner, C.L. 1985. The ecology of wolverines in southcentral Alaska. M.S. thesis, University of Alaska, Fairbanks, Alaska.

Garrett, M.G., J.W. Watson, and R.G. Anthony. 1993. Bald Eagle home range and habitat use in the Columbia River estuary. Journal of Wildlife Management 57(1):19-27.

Gilbert, F.F., and R. Allwine. 1991a. Spring bird communities in the Oregon Cascade Range. Pages 145-158 in L.F. Ruggiero, K.B. Aubry, A.B. Carey, and M.H. Huff (eds.). Wildlife and vegetation of unmanaged Douglas-fir forests. USDA Forest Service General Technical Report PNW-GTR-285. Portland, Ore.

–. 1991b. Small mammal communities in the Oregon Cascade Range. Pages 257-267 in L.F. Ruggiero, K.B. Aubry, A.B. Carey, and M.H. Huff (eds.). Wildlife and vegetation of unmanaged Douglas-fir forests. USDA Forest Service General Technical Report PNW-GTR-285. Portland, Ore.

–. 1991c. Terrestrial amphibian communities in the Oregon Cascade Range. Pages 319-324 in L.F. Ruggiero, K.B. Aubry, A.B. Carey, and M.H. Huff (eds.). Wildlife and vegetation of unmanaged Douglas-fir forests. USDA Forest Service General Technical Report PNW-GTR-285. Portland, Ore.

Goward, T. 1993a. Epiphytic lichens: going down with the trees. Pages 153-158 in S. Rautio (ed.). Community action for endangered species: a public symposium on B.C.'s threatened and endangered species and their habitats. Federation of B.C. Naturalists, Vancouver, B.C.

–. 1993b. Crown of the ICH: epiphytic macrolichens of old growth forests in the Interior Cedar-Hemlock zone. BioLine 11(2):15-17.

–. 1994a. Living antiquities. Nature Canada (Summer):15-21.

–. 1994b. Notes on oldgrowth-dependent epiphytic macrolichens in inland British Columbia, Canada. Acta Botany Fennica 150:31-38.

Goward, T., and B. McCune. (in prep.). The lichens of British Columbia. Illustrated keys. Part 2. Fruticose species. Special Report Series 20. B.C. Ministry of Forests, Research Branch, Victoria, B.C.

Goward, T., B. McCune, and D. Meidinger. 1994. The lichens of British Columbia. Illustrated keys. Part 1. Foliose and squamulose species. Special Report Series 8. B.C. Ministry of Forests, Research Branch, Victoria, B.C.

Gustafsson, L., A. Fiskesjö, T. Hallingbäck, T. Ingelög, and B. Pettersson. 1992a. Semi-natural deciduous broadleaved woods in southern Sweden – habitat factors of importance to some bryophyte species. Biological Conservation 59:175-181.

Gustafsson, L., A. Fiskesjö, T. Ingelög, B. Pettersson, and G. Thor. 1992b. Factors of importance to some lichen species of deciduous broad-leaved woods in southern Sweden. Lichenologist 24(3):255-266.

Habeck, J.R. 1968. Forest succession in the Glacier Park cedar-hemlock forests. Ecology 49(5):872-880.

–. 1978. A study of climax western redcedar (*Thuja plicata* Donn.) forest communities in the Selway-Bitterroot Wilderness, Idaho. Northwest Science 52(1):67-76.

–. 1988. Old-growth forests in the northern Rocky Mountains. Natural Areas Journal 8(3):202-211.

–. 1990. Old-growth Ponderosa pine–western larch forests in western Montana: ecology and management. Northwest Environmental Journal 6:271-292.

Halpern, C.B. 1988. Early successional pathways and the resistance and resilience of forest communities. Ecology 69(6):1703-1715.

Halpern, C.B., and T.A. Spies. 1995. Plant species diversity in natural and managed forests of the Pacific Northwest. Ecological Applications 5(4):913-934.

Hanley, T.A., R.G. Cates, B. Van Horne, and J.D. McKendrick. 1987. Forest stand-age-related differences in apparent nutritional quality of forage for deer in southeastern Alaska. Pages 9-17 in F.D. Provenza, J.T. Flinders, and E.D. McArthur. Proceedings – symposium on plant-herbivore interactions. USDA Forest Service General Technical Report INT-222. Ogden, Utah.

Harding, L. 1994. Threats to diversity of forest ecosystems in British Columbia. Pages 245-277 in L.E. Harding, and E. McCullum (eds.). Biodiversity in British Columbia. Our changing environment. Environment Canada/Canadian Wildlife Service, Ottawa, Ont.

Harestad, A.S., and D.G. Keisker. 1989. Nest tree use by primary cavity-nesting birds in south-central British Columbia. Canadian Journal of Zoology 67:1067-1073.

Hatler, D.F., R.W. Campbell, and A. Dorst. 1978. Birds of Pacific Rim Park, British Columbia. Occasional Paper Number 20. B.C. Provincial Museum, Victoria, B.C.

Holroyd, S.L., R.M.R. Barclay, L.M. Merk, and R.M. Brigham. 1994. A survey of the bat fauna of the dry interior of British Columbia. Wildlife Working Report No. WR-63. B.C. Ministry of Environment, Lands and Parks, Wildlife Branch, Victoria, B.C.

Hornocker, M.G., and H.S. Hash. 1981. Ecology of the wolverine in northwestern Montana. Canadian Journal of Zoology 59:1286-1301.

Howie, R., and R. Ritcey. 1987. Distribution, habitat selection, and densities of Flammulated Owls in British Columbia. Pages 249-254 in R.W. Nero, R.J. Clark, R.J. Knapton, and R.H. Hamre (eds.). Biology and conservation of Northern Forest Owls: symposium proceedings. USDA Forest Service General Technical Report RM-142. Rocky Mountain Forest and Range Experimental Station, Fort Collins, Colo.

Huff, M.H., and C.M. Raley. 1991. Regional patterns of diurnal breeding bird communities in Oregon and Washington. Pages 177-205 in L.F. Ruggiero, K.B. Aubry, A.B. Carey, and M.H. Huff (eds.). Wildlife and vegetation of unmanaged Douglas-fir forests. USDA Forest Service General Technical Report PNW-GTR-285. Portland, Ore.

Huff, M.H., D.A. Manuwal, and J.A. Putera. 1991. Winter bird communities in the southern Washington Cascade Range. Pages 207-218 in L.F. Ruggiero, K.B. Aubry, A.B. Carey, and M.H. Huff (eds.). Wildlife and vegetation of unmanaged Douglas-fir forests. USDA Forest Service General Technical Report PNW-GTR-285. Portland, Ore.

Isaacs, F.B., R. Goggans, R.G. Anthony, and T. Bryan. 1993. Habits of Bald Eagles wintering along the Crooked River, Oregon. Northwest Science 67(2):55-62.

Johnson, E.A., and G.I. Fryer. 1989. Population dynamics in lodgepole pine–Engelmann spruce forests. Ecology 70(5):1335-1345.

Keisker, D.G. 1987. Nest tree selection by primary cavity-nesting birds in south-central British Columbia. Wildlife Report Number R-13. B.C. Ministry of Environment, Wildlife Branch, Victoria, B.C.

Klinka, K., A.M. Scagel, and P.J. Courtin. 1985. Vegetation relationships among some seral ecosystems in southwestern British Columbia. Canadian Journal of Forest Research 15:561-569.

Kneeshaw, D.D. 1992. Tree population dynamics of some old sub-boreal spruce stands. M.Sc. thesis, University of British Columbia, Faculty of Forestry, Vancouver, B.C.

Koehler, G.M., and K.B. Aubry. 1994. Lynx. Pages 74-98 in L.F. Ruggiero, K.B. Aubry, S.W. Buskirk, L.J. Lyon, and W.J. Zielinski (tech. eds.). The scientific basis for conserving forest carnivores. American marten, fisher, lynx, and wolverine in the western United States. USDA Forest Service General Technical Report RM-254. Fort Collins, Colo. 184 pp.

Koehler, G.M., and J.D. Brittell. 1990. Managing spruce-fir habitat for lynx and snowshoe hares. Journal of Forestry 88:10-14.

Koehler, G.M., J.A. Blakesley, and T.W. Koehler. 1990. Marten use of successional forest stages during winter in north-central Washington. Northwest Naturalist 71:1-4.

Lance, A.N., and M. Phinney. 1993. Bird diversity and abundance following aspen clearcutting in the Boreal White and Black Spruce biogeoclimatic zone. Report for 1992, Project B28, Hardwood and Vegetation Management Research. B.C. Ministry of Forests, Research Branch, Victoria, B.C.

Lattin, J.D., and A.R. Moldenke. 1990. Moss lacebugs in northwest conifer forests: adaptation to long-term stability. Northwest Environmental Journal 6(2):406-407.

Lee, P.C., S. Crites, and J.B. Stelfox. 1995a. Changes in forest structure and floral composition in a chronosequence of aspen mixedwood stands in Alberta. Pages 29-48 in J.B. Stelfox (ed.). Relationships between stand age, stand structure, and biodiversity in aspen mixedwood forests in Alberta. Jointly published by Alberta Environmental Centre (AECV95-R1), Vegreville, Alta., and Canadian Forest Service (Project No. 0001A), Edmonton, Alta. 308 pp.

Lee, P.C., S. Crites, M. Nietfeld, H.V. Nguyen, and J.B. Stelfox. 1995b. Changes in snags and down woody material characteristics in a chronosequence of aspen mixedwood forests in Alberta. Pages 49-61 in J.B. Stelfox (ed.). Relationships between stand age, stand structure, and biodiversity in aspen mixedwood forests in Alberta. Jointly published by Alberta Environmental Centre (AECV95-R1), Vegreville, Alta., and Canadian Forest Service (Project No. 0001A), Edmonton, Alta. 308 pp.

Lee, P.C., S. Crites, K. Sturgess, and J.B. Stelfox. 1995c. Changes in understory composition for a chronosequence of aspen mixedwood stands in Alberta. Pages 63-90 in J.B. Stelfox (ed.). Relationships between stand age, stand structure, and biodiversity in aspen mixedwood forests in Alberta. Jointly published by Alberta Environmental Centre (AECV95-R1), Vegreville, Alta., and Canadian Forest Service (Project No. 0001A), Edmonton, Alta. 308 pp.

Lertzman, K.P. 1992. Patterns of gap-phase replacement in a subalpine, old-growth forest. Ecology 73:657-669.

Lertzman, K.P., and C.J. Krebs. 1991. Gap-phase structure of a subalpine old-growth forest. Canadian Journal of Forestry Research 21:1730-1741.

Lesica, P., B. McCune, S.V. Cooper, and W.S. Hong. 1991. Differences in lichen and bryophyte communities between old-growth and managed second-growth forests in the Swan Valley, Montana. Canadian Journal of Botany 69:1745-1755.

Lundquist, R.W., and J.M. Mariani. 1991. Nesting habitat and abundance of snag-dependent birds in the southern Washington Cascade Range. Pages 221-240 in L.F. Ruggiero, K.B. Aubry, A.B. Carey, and M.H. Huff (eds.). Wildlife and vegetation of unmanaged Douglas-fir forests. USDA Forest Service General Technical Report PNW-GTR-285. Portland, Ore.

Luoma, D.L. 1991. Annual changes in seasonal production of hypogeous sporocarps in Oregon Douglas-fir forests. Pages 83-89 in L.F. Ruggiero, K.B. Aubry, A.B. Carey, and M.H. Huff (eds.). Wildlife and vegetation of unmanaged Douglas-fir forests. USDA Forest Service General Technical Report PNW-GTR-285. Portland, Ore.

Luoma, D.L., and R.E. Frenkel. 1991. Fruiting of hypogeous fungi in Oregon Douglas-fir forests: seasonal and habitat variation. Mycologia 83(3):335-353.

MacKinnon, A., and M. Eng. 1995. Old forests inventory for coastal British Columbia. Cordillera 2(1):20-33.

Manuwal, D.A. 1991. Spring bird communities in the southern Washington Cascade Range. Pages 161-174 in L.F. Ruggiero, K.B. Aubry, A.B. Carey, and M.H. Huff (eds.). Wildlife and vegetation of unmanaged Douglas-fir forests. USDA Forest Service General Technical Report PNW-GTR-285. Portland, Ore.

Manuwal, D.A., and M.H. Huff. 1987. Spring and winter bird populations in a Douglas-fir forest sere. Journal of Wildlife Management 51(3):586-595.

Marshall, D.B. 1992. Status of the Black-backed Woodpecker in Oregon and Washington. Audubon Society of Portland, 26-29 May 1992. Blue Mountains Biodiversity Conference, Portland, Ore.

Marshall, V.G. 1993. Sustainable forestry and soil fauna diversity. Pages 239-248 in M.A. Fenger, E.H. Miller, J.F. Johnson, and E.J.R. Williams (eds.). Our living legacy. Proceedings of a symposium on biological diversity. Royal British Columbia Museum, Victoria, B.C.

Marshall, V.G., D.K.M. Kevan, J.V. Matthews Jr., and A.D. Tomlin. 1982. Status and research needs of Canadian soil fauna. Bulletin of the Entomological Society of Canada, Supplement 14.

Maser, C., J.M. Trappe, and R.A. Nussbaum. 1978. Fungal–small mammal interrelationships with emphasis on Oregon coniferous forests. Ecology 59(4):799-809.

McCune, B., and J.A. Antos. 1981a. Correlations between forest layers in the Swan Valley, Montana. Ecology 62(5):1196-1204.

–. 1981b. Diversity relationships of forest layers in the Swan Valley, Montana. Bulletin of the Torrey Botanical Club 108:354-361.

–. 1982. Epiphyte communities of the Swan Valley, Montana. Bryologist 85:1-12.

McDonald, L. 1995. Relationships between northern flying squirrels and stand age and structure in aspen mixedwood forests in Alberta. Pages 227-240 in J.B. Stelfox (ed.). Relationships between stand age, stand structure, and biodiversity in aspen mixedwood forests in Alberta. Jointly published by Alberta Environmental Centre (AECV95-R1), Vegreville, Alta., and Canadian Forest Service (Project No. 0001A), Edmonton, Alta. 308 pp.

Milne, K., and S. Hejl. 1989. Nest site characteristics of White-headed Woodpeckers. Journal of Wildlife Management 53(1):50-55.

Moeur, M. 1992. Baseline demographics of late successional western hemlock/western redcedar stands in northern Idaho Research Natural Areas. USDA Forest Service Research Paper INT-456. Ogden, Utah.

Moldenke, A.R., and J.D. Lattin. 1990a. Dispersal characteristics of old-growth soil arthropods: the potential for loss of diversity and biological function. Northwest Environmental Journal 6(2):408-409.

–. 1990b. Density and diversity of soil arthropods as 'biological probes' of complex soil phenomena. Northwest Environmental Journal 6(2):409-410.

Molina, R., H. Massicotte, and J.M. Trappe. 1992. Specificity phenomena in mycorrhizal symbiosis: community ecological consequences and practical implications. Pages 357-423 in M. Allen (ed.). Mycorrhizal functioning. Chapman and Hall, New York.

Nagorsen, D.W., and R.M. Brigham. 1993. The mammals of British Columbia. Volume 1. The bats of British Columbia. Royal British Columbia Museum, Victoria, B.C., and UBC Press, Vancouver, B.C.

Nehlsen, W., J.E. Williams, and J.A. Lichatowich. 1991. Pacific salmon at the crossroads: stocks at risk from California, Oregon, Idaho and Washington. Fisheries 16(2):4-21.

Nyberg, J.B., and D.W. Janz (eds.). 1990. Deer and elk habitats in coastal forests of southern Vancouver Island. Special Report Series No. 5. B.C. Ministry of Forests and B.C. Ministry of Environment, Victoria, B.C.

Nyberg, J.B., F.L. Bunnell, D.W. Janz, and R.M. Ellis. 1986. Managing young forests as black-tailed deer winter ranges. Land Management Report No. 37. B.C. Ministry of Forests, Victoria, B.C.

Nyberg, J.B., A.S. Harestad, and F.L. Bunnell. 1987. 'Old growth' by design: managing young forests for old-growth wildlife. Pages 70-81 in Transactions of the 52nd North American

Wildlife and Natural Resources Conference, 1987, Quebec City. Wildlife Management Institute, Washington, D.C.

Pauley, G.R., J.M. Peek, and P. Zager. 1993. Predicting white-tailed deer habitat use in northern Idaho. Journal of Wildlife Management 57(4):904-913.

Perkins, J.M., and S.P. Cross. 1988. Differential use of some coniferous forest habitats by hoary and silver-haired bats in Oregon. Murrelet 69(1):21-24.

Perry, D.A., R. Molina, and M.P. Amaranthus. 1987. Mycorrhizae, mycorrhizospheres, and reforestation: current knowledge and research needs. Canadian Journal of Forest Research 17:929-940.

Pike, L.H., R.A. Rydell, and W.C. Denison. 1977. A 400-year-old Douglas fir tree and its epiphytes: biomass, surface area, and their distributions. Canadian Journal of Forest Research 7:680-699.

Pike, L.H., D.M. Tracy, M.A. Sherwood, and D. Nielsen. 1972. Estimates of biomass and fixed nitrogen of epiphytes from old-growth Douglas-fir. Pages 177-187 in J.F. Franklin, L.J. Dempster, and R.H. Waring (eds.). Research on coniferous forest ecosystems. USDA Forest Service, Portland, Ore.

Pojar, J. 1993. Terrestrial diversity of British Columbia. Pages 177-190 in M.A. Fenger, E.H. Miller, J.F. Johnson, and E.J.R. Williams (eds.). Our living legacy. Proceedings of a symposium on biological diversity. Royal British Columbia Museum, Victoria, B.C.

Pojar, J., and A. MacKinnon (eds.). 1994. Plants of coastal British Columbia. Lone Pine Publishing, Edmonton, Alta.

Pojar, J., E. Hamilton, D. Meidinger, and A. Nicholson. 1992. Old growth forests and biological diversity in British Columbia. Pages 85-97 in G.B. Ingram and M.R. Moss (eds.). Landscape approaches to wildlife and ecosystem management. Proceedings of the Second Symposium of the Canadian Society for Landscape Ecology and Management. Polyscience Publications Inc., Morin Heights, Ont.

Pough, F.H., E.M. Smith, D.H. Rhodes, and A. Collazo. 1987. The abundance of salamanders in forest stands with different histories of disturbance. Forest Ecology and Management 20:1-9.

Powell, R.A., and W.J. Zielinski. 1994. Fisher. Pages 38-73 in L.F. Ruggiero, K.B. Aubry, S.W. Buskirk, L.J. Lyon, and W.J. Zielinski (tech. eds.). The scientific basis for conserving forest carnivores. American marten, fisher, lynx, and wolverine in the western United States. USDA Forest Service General Technical Report RM-254. Fort Collins, Colo. 184 pp.

Radcliffe, G., B. Bancroft, G. Porter, and C. Cadrin. 1994. Biodiversity of the Prince Rupert Forest Region. Land Management Report No. 82. B.C. Ministry of Forests, Research Branch, Victoria, B.C.

Rainey, W.E., E.D. Pierson, M. Colberg, and J.H. Barclay. 1992. Bats in hollow redwoods: seasonal use and role in nutrient transfer into old growth communities. Bat Research News 33(4):71.

Ralph, C.J., P.W.C. Paton, and C.A. Taylor. 1991. Habitat association patterns of breeding birds and small mammals in Douglas-fir/hardwood stands in northwestern California and southwestern Oregon. Pages 379-393 in L.F. Ruggiero, K.B. Aubry, A.B. Carey, and M.H. Huff (eds.). Wildlife and vegetation of unmanaged Douglas-fir forests. USDA Forest Service General Technical Report PNW-GTR-285. Portland, Ore.

Raphael, M.G. 1988. Long-term trends in abundance of amphibians, reptiles, and mammals in Douglas-fir forests in northwestern California. Pages 23-31 in R.C. Szaro, K.E. Severson, and D.R. Patton (tech. coords.). Management of amphibians, reptiles and small mammals in North America. USDA Forest Service General Technical Report RM-166. Fort Collins, Colo.

–. 1991. Vertebrate species richness within and among seral stages of Douglas-fir/hardwood forest in northwestern California. Pages 415-423 in L.F. Ruggiero, K.B. Aubry, A.B. Carey, and M.H. Huff (eds.). Wildlife and vegetation of unmanaged Douglas-fir forests. USDA Forest Service General Technical Report PNW-GTR-285. Portland, Ore.

Raphael, M.G., and B.G. Marcot. 1986. Validation of a wildlife-habitat-relationships model: vertebrates in Douglas-fir sere. Pages 129-138 in J. Verner, M.L. Morrison, and C.J. Ralph (eds.). Wildlife 2000: modeling habitat relationships of terrestrial vertebrates. University of Wisconsin Press, Madison, Wis.

Raphael, M.G., and M. White. 1984. Use of snags by cavity-nesting birds in the Sierra Nevada. Wildlife Monographs 86:1-66.

Redhead, S. 1994. Macrofungi of British Columbia. Pages 81-89 in L.E. Harding, and E. McCullum (eds.). Biodiversity in British Columbia. Our changing environment. Environment Canada/Canadian Wildlife Service, Ottawa, Ont.

Redhead, S., and S. Berch. 1997. Standardized inventory methodologies for components of British Columbia's biodiversity: macrofungi. Rsources Inventory Committee, Victoria, B.C.

Redhead, S.A., and L.L. Norvell. 1993. *Phaeocollybia* in western Canada. Mycotaxon 46:343-358.

Rodway, M.S., H.R. Carter, S.G. Sealy, and R.W. Campbell. 1992. Status of the Marbled Murrelet in British Columbia. Pages 17-41 in H.R. Carter and M.L. Morrison (eds.). Status and conservation of the Marbled Murrelet in North America. Proceedings of the Western Foundation of Vertebrate Zoology 5(1).

Rodway, M.S., H.M. Regehr, and J.-P. Savard. 1993. Activity patterns of Marbled Murrelets in old-growth forest in the Queen Charlotte Islands, British Columbia. Condor 95:831-848.

Roe, N.A. 1974. Birds and disturbed forest succession after logging in Pacific Rim National Park, Vancouver Island, British Columbia. M.Sc. thesis, University of Calgary, Calgary, Alta.

Rominger, E.M., and J.L. Oldemeyer. 1989. Early-winter habitat of woodland caribou, Selkirk Mountains, British Columbia. Journal of Wildlife Management 53(1):238-243.

Rosenberg, D.K., and R.G. Anthony. 1992. Characteristics of northern flying squirrel populations in young second- and old-growth forests in western Oregon. Canadian Journal of Zoology 70:161-166.

Rosenberg, K.V., and M.G. Raphael. 1986. Effects of forest fragmentation on vertebrates in Douglas-fir forests. Pages 263-272 in J. Verner, M.L. Morrison, and C.J. Ralph (eds.). Wildlife 2000: modeling habitat relationships of terrestrial vertebrates. University of Wisconsin Press, Madison, Wis.

Roy, L.D., J.B. Stelfox, and J.W. Nolan. 1995. Relationships between mammal biodiversity and stand age and structure in aspen mixedwood forests in Alberta. Pages 159-189 in J.B. Stelfox (ed.). Relationships between stand age, stand structure, and biodiversity in aspen mixedwood forests in Alberta. Jointly published by Alberta Environmental Centre (AECV95-R1), Vegreville, Alta., and Canadian Forest Service (Project No. 0001A), Edmonton, Alta. 308 pp.

Ruggiero, L.F. 1991. Wildlife habitat relationships and viable populations. Pages 443-445 in L.F. Ruggiero, K.B. Aubry, A.B. Carey, and M.H. Huff (eds.). Wildlife and vegetation of unmanaged Douglas-fir forests. USDA Forest Service General Technical Report PNW-GTR-285. Portland, Ore.

Ruggiero, L.F., and A.B. Carey. 1984. A programmatic approach to the study of old-growth forest-wildlife relationships. Pages 340-345 in Proceedings of the annual convention of the Society of American Foresters, Portland, Oregon. Society of American Foresters, Bethesda, Md.

Ruggiero, L.F., R.S. Holthausen, B.G. Marcot, K. Aubry, J.W. Thomas, and E.C. Meslow. 1988. Ecological dependency: the concept and its implications for research and management. Pages 115-126 in Transactions of the 53rd North American Wildlife and Natural Resources Conference, 1988 Wildlife Management Institute, Washington, D.C.

Ruggiero, L.F., L.L.C. Jones, and K.B. Aubry. 1991. Plant and animal habitat associations in Douglas-fir forests of the Pacific Northwest: an overview. Pages 447-462 in L.F. Ruggiero, K.B. Aubry, A.B. Carey, and M.H. Huff (eds.). Wildlife and vegetation of unmanaged Douglas-fir forests. USDA Forest Service General Technical Report PNW-GTR-285. Portland, Ore.

Savard, J.-P., D.A. Seip, and L. Waterhouse. (in prep.). Forest birds in relation to age and structure of coastal forests in British Columbia. Regional Technical Report, Canadian Wildlife Service, Pacific & Yukon Region, Delta, B.C.

Schieck, J., M. Nietfeld, and J.B. Stelfox. 1995. Differences in bird species richness and abundance among three successional stages of aspen-dominated boreal forests. Canadian Journal of Zoology 73(8):1417-1431.

Schoen, J.W., M.D. Kirchhoff, and J.H. Hughes. 1988. Wildlife and old-growth forests in southeastern Alaska. Natural Areas Journal 8(3):138-145.

Schoen, J.W., M.D. Kirchhoff, and O.C. Wallmo. 1984. Sitka black-tailed deer/old-growth relationships in southeast Alaska: implications for management. Pages 315-319 in W.R. Meehan, T.R. Merrell Jr., and T.A. Handley (eds.). Fish and wildlife relationships in old-growth forests. Proceedings of a symposium. 12-15 April 1982, Juneau, Alaska. American Institute of Fishery Research Biologists, Moorehead City, N.C. 425 pp.

Schoonmaker, P., and A. McKee. 1988. Species composition and diversity during secondary succession of coniferous forests in the western Cascade Mountains of Oregon. Forest Science 34(4):960-979.

Schowalter, T.D. 1989. Canopy arthropod community structure and herbivory in old-growth and regenerating forests in western Oregon. Canadian Journal of Forest Research 19:318-322.

Scudder, G.G.E. 1994. An annotated systematic list of the potentially rare and endangered freshwater and terrestrial invertebrates in British Columbia. Entomological Society of British Columbia Occasional Paper 2:1-92.

Sealy, S.G., and H.R. Carter. 1984. At-sea distribution and nesting habitat of the Marbled Murrelet in British Columbia: problems in the conservation of a solitary nesting seabird. Pages 737-756 in J.P. Croxall, P.G.H. Evans, and R.W. Schreiber (eds.). Status and conservation of the world's seabirds. Technical Publication No. 2. International Council for Bird Preservation.

Seip, D. 1996. The projected impacts of different biodiversity emphasis options on some forest bird species in the Sub-Boreal (SBS) Zone. Note #PG-04, B.C. Ministry of Forests, Prince George, B.C.

Seip, D. (in review). Abundnce of birds and small mammals in relation to the age and structure of coniferous forests in central interior British Columbia. Submitted to Canadian Journal of Forest Research.

Seip, D., and J.-P. Savard. (in prep.). Wildlife diversity in old growth forests and managed stands. B.C. Ministry of Forests, Prince George, B.C.

Selva, S.B. 1994. Lichen diversity and stand continuity in northern hardwoods and spruce-fir forests of northern New England and western New Brunswick. Bryologist 97:424-429.

Setälä, H., and V. Marshall. 1994. Stumps as a habitat for Collembola during succession from clear-cuts to old-growth Douglas-fir forests. Pedobiologia 38:307-326.

Soderstrom, L. 1988. The occurrence of epixylic bryophyte and lichen species in an old natural and a managed forest stand in northeast Sweden. Biological Conservation 45:169-178.

–. 1989. Regional distribution patterns of bryophyte species on spruce logs in northern Sweden. Bryologist 92:349-355.

Spies, T.A. 1991a. Plant species diversity and occurrence in young, mature, and old-growth Douglas-fir stands in western Oregon and Washington. Pages 111-121 in L.F. Ruggiero, K.B. Aubry, A.B. Carey, and M.H. Huff (eds.). Wildlife and vegetation of unmanaged Douglas-fir forests. USDA Forest Service General Technical Report PNW-GTR-285. Portland, Ore.

–. 1991b. Current knowledge of old growth in the Douglas-fir region of western North America. Pages 22-33 in Proceedings, Twelfth Annual Forest Vegetation Management Conference, Redding, Ca.

Spies, T.A., and S.P. Cline. 1988. Coarse woody debris in forests and plantations of coastal Oregon. Pages 5-23 in C. Maser, R.F. Tarrant, J.M. Trappe, and J.F. Franklin (tech eds.). From the forest to the sea: a story of fallen trees. USDA Forest Service General Technical Report PNW-GTR-229. Published in cooperation with the USDI Bureau of Land Management. Portland, Ore. 153 pp.

Spies, T.A., and J.F. Franklin. 1991. The structure of natural young, mature, and old-growth Douglas-fir forests in Oregon and Washington. Pages 91-109 in L.F. Ruggiero, K.B. Aubry,

A.B. Carey, and M.H. Huff (eds.). Wildlife and vegetation of unmanaged Douglas-fir forests. USDA Forest Service General Technical Report PNW-GTR-285. Portland, Ore.

Stelfox, J.B., L.D. Roy, and J. Nolan. 1995. Abundance of ungulates in relation to stand age and structure in aspen mixedwood forests in Alberta. Pages 181-200 in J.B. Stelfox (ed.). Relationships between stand age, stand structure, and biodiversity in mixedwood forests in Alberta. Jointly published by Alberta Environmental Centre (AECV95-R1), Vegreville, Alta., and Canadian Forest Service (Project No. 0001A), Edmonton, Alta.

Stelmock, J.J., and A.S. Harestad. 1979. Food habits and life history of the clouded salamander (*Aneides ferreus*) on northern Vancouver Island, British Columbia. Syesis 12:71-75.

Sullivan, T.P., and C.J. Krebs. 1980. Comparative demography of *Peromyscus maniculatus* and *Microtus oregoni* populations after logging and burning of coastal forest habitats. Canadian Journal of Zoology 58:2252-2259.

Terry, C.J. 1981. Habitat differentiation among three species of *Sorex* and *Neurotrichus gibbsi* in Washington. American Midland Naturalist 106:119-125.

Thomas, D.W. 1988. The distribution of bats in different ages of Douglas-fir forests. Journal of Wildlife Management 52(4):619-626.

Thomas, D.W., and S.D. West. 1991. Forest age associations of bats in the southern Washington Cascade and Oregon Coast Ranges. Pages 295-303 in L.F. Ruggiero, K.B. Aubry, A.B. Carey, and M.H. Huff (eds.). Wildlife and vegetation of unmanaged Douglas-fir forests. USDA Forest Service General Technical Report PNW-GTR-285. Portland, Ore.

Thomas, J.W. (tech. ed.). 1979. Wildlife habitats in managed forests: the Blue Mountains of Oregon and Washington. Agriculture Handbook No. 553. USDA Forest Service Agricultural Handbook No. 533. Portland, Ore. 512 pp.

Thomas, J.W., M.G. Raphael, R.G. Anthony, E.D. Forsman, A.G. Gunderson, R.S. Holthausen, B.G. Marcot, G.H. Reeves, J.R. Sedell, and D.M. Solis. 1993. Viability assessments and management considerations for species associated with late-successional and old-growth forests of the Pacific Northwest. The report of the Scientific Analysis Team. USDA Forest Service, Portland, Ore.

Thompson, I.D. 1994. Marten populations in uncut and logged boreal forests in Ontario. Journal of Wildlife Management 58(2):272-280.

Thompson, I.D., and P.W. Colgan. 1994. Marten activity in uncut and logged boreal forests in Ontario. Journal of Wildlife Management 58(2):280-288.

Thompson, I.D., and A.S. Harestad. 1994. Effects of logging on American martens, and models for habitat management. Pages 355-367 in S.W. Buskirk, A.S. Harestad, M.G. Raphael, and R.A. Powell (eds.). Martens, sables, and fishers: biology and conservation. Cornell University Press, Ithaca, N.Y. 484 pp.

Turner, D.P., and E.H. Franz. 1985. Size class structure and tree dispersion patterns in old-growth cedar-hemlock forests of the northern Rocky Mountains (USA). Oecologia 68:52-56.

Ure, D.C., and C. Maser. 1982. Mycophagy of red-backed voles in Oregon and Washington. Canadian Journal of Zoology 60:3307-3315.

USDA Forest Service and USDI Bureau of Land Management. 1994. Supplemental Environmental Impact Statement (SEIS) Interdisciplinary Team. Final supplemental environmental impact statement on management of habitat for late-successional and old-growth forest related species within the range of the Northern Spotted Owl. 2 volumes + appendices. Portland, Ore.

Vermeer, K., and K.H. Morgan. 1989. Nesting population, nest sites, and prey remains of Bald Eagles in Barkley Sound. Northwest Science 70:21-26.

Vermeer, K., K.H. Morgan, R.W. Butler, and G.E.J. Smith. 1989. Population, nesting habitat, and food of Bald Eagles in the Gulf Islands. Pages 123-130 in K. Vermeer et al. (eds.). The ecology and status of marine and shoreline birds in the Strait of Georgia, British Columbia. Environment Canada/Canadian Wildlife Service, Ottawa, Ont.

Walls, S.C., A.R. Blaustein, and J.J. Beatty. 1992. Amphibian biodiversity of the Pacific Northwest with special reference to old-growth stands. Northwest Environmental Journal 8:53-69.

Welsh, H.H. Jr. 1990. Relictual amphibians and old-growth forests. Conservation Biology 4(3):309-319.

Welsh, H.H. Jr., and A.J. Lind. 1988. Old growth forests and the distribution of the terrestrial herpetofauna. Pages 439-458 in R.C. Szaro, K.E. Severson, and D.R. Patton (tech. coords.). Management of amphibians, reptiles and small mammals in North America. USDA Forest Service General Technical Report RM-166. Fort Collins, Colo.

–. 1991. The structure of the herpetofaunal assemblage in the Douglas-fir/hardwood forests of northwestern California and southwestern Oregon. Pages 395-413 in L.F. Ruggiero, K.B. Aubry, A.B. Carey, and M.H. Huff (eds.). Wildlife and vegetation of unmanaged Douglas-fir forests. USDA Forest Service General Technical Report PNW-GTR-285. Portland, Ore.

West, S.D. 1991. Small mammal communities in the southern Washington Cascade Range. Pages 269-283 in L.F. Ruggiero, K.B. Aubry, A.B. Carey, and M.H. Huff (eds.). Wildlife and vegetation of unmanaged Douglas-fir forests. USDA Forest Service General Technical Report PNW-GTR-285. Portland, Ore.

Westworth, D.A., and E.S. Telfer. 1993. Summer and winter bird populations associated with five age-classes of aspen forest in Alberta. Canadian Journal of Forest Research 23:1830-1836.

Wetmore, S.P., R.A. Keller, and G.E. John Smith. 1985. Effects of logging on bird populations in British Columbia as determined by a modified point-count method. Canadian Field Naturalist 99(2):224-233.

Whitaker, J.O. Jr., C. Maser, R.M. Storm, and J.J. Beatty. 1986. Food habits of clouded salamanders (*Aneides ferreus*) in Curry County, Oregon (Amphibia:Caudata:Plethodontidae). Great Basin Naturalist 46:228-240.

Winchester, N.N. (1997). Canopy arthropods of coastal Sitka spruce trees on Vancouver Island, British Columbia, Canada. Pages 151-168 in N.E. Stork, J.A. Adis, and R.K. Didham (eds.). Canopy Arthropods. Chapman and Hall, London, England.

Winchester, N.N., and R.A. Ring. (1996). Centinelan extinctions: extirpation of northern temperate old-growth rainforest arthropod communities. Selybana 17(1):50-57.

Witt, J.W. 1992. Home range and density estimates for the northern flying squirrel, *Glaucomys sabrinus,* in western Oregon. Journal of Mammalogy 73(4):921-929.

Zarnowitz, J.E., and D.A. Manuwal. 1985. The effects of forest management on cavity-nesting birds in northwestern Washington. Journal of Wildlife Management 49:255-263.

Zuleta, G.A., and C. Galindo-Leal. 1994. Distribution and abundance of four species of small mammals at risk in a fragmented landscape. Wildlife Working Report No. WR-64. B.C. Ministry of Environment, Lands and Parks, Wildlife Branch, Victoria, B.C.

7
The Dead Wood Cycle
Eric Lofroth

Definitions

Dead wood cycle: This term is rarely used in the literature. It refers to the process of tree death, tree fall, and decay in the forested ecosystem. The cycle begins with live healthy trees and ends with their incorporation into the soil organic horizon and/or aquatic environment.

Coarse woody debris: This term has been defined variously by researchers and managers. Some (e.g., Harmon et al. 1986) have used it to describe all states of dead wood in the cycle, from snags to logs and fallen branches. Others (Lofroth 1993; Steventon 1994; Province of British Columbia 1995) define it as downed woody material, distinguishing it from the 'snag' or standing dead component. The size above which debris is considered 'coarse' varies among studies and has been identified as greater than 2.5 cm in diameter (Harmon et al. 1986), greater than 5 cm in diameter (Mattson et al. 1987), greater than 7.5 cm in diameter (Quesnel 1994; Keenan and Inselberg, in prep.), greater than 10 cm in diameter (Spies et al. 1988), greater than 15 cm in diameter (Sollins 1982), and greater than 20 cm in diameter (Harmon et al. 1987).

Here coarse woody debris (CWD) is defined as downed woody material greater than 10 cm in diameter, and is arbitrarily distinguished from standing dead woody material (wildlife trees or snags) by the angle of repose (less than 45° from the ground).

Snags: dead standing trees, typically with a specified lower size limit. Cline et al. (1980) sampled snags greater than 9 cm in diameter at breast height (dbh) and greater than 4.4 m tall. Raphael and Morrison (1987) set lower boundaries on snags of 13 cm dbh and 1.5 m tall. Lofroth (1993) set a lower limit of 7.5 cm dbh and 2 m tall. The B.C. Ministry of Forests and B.C. Workers' Compensation Board define a snag as a standing dead tree greater than 3 m in height (Backhouse and Lousier 1991). Here snags are considered to be standing dead trees greater than 10 cm dbh and greater than 2 m tall, with an angle of repose greater than 45° from the ground.

Wildlife tree: 'a tree that provides present or future valuable habitat for the conservation or enhancement of wildlife' (Guy and Manning 1995). Wildlife trees may be distinguished by attributes such as structure, age, abundance, location, and surrounding habitat features. They range from live and healthy trees to decayed stubs, and, as such, include snags.

Large organic debris: essentially, coarse woody debris in aquatic ecosystems. Large organic debris (LOD) is downed woody material, including tree boles, limbs, and rooting structures. Van Sickle and Gregory (1990) define it as material greater than 10 cm in diameter and greater than 1.5 m long. Bisson et al. (1987) used a limit of 10 cm in diameter but did not specify a minimum length. Here LOD is defined as woody material in aquatic ecosystems greater than 10 cm in diameter.

Fine fuel/slash: the debris left behind following forest harvesting, including woody material of any size (McRae et al. 1979; Trowbridge et al. 1987).

Dead wood obligate: organism that *requires* some component of the dead wood cycle for some or all of its life history. The conservation of populations of such species can be considered dependent on the presence of dead wood.

Dead wood facultative: organism that *may use but does not require* some component of the dead wood cycle for some or all of its life history.

Background

Dead and downed woody debris in western forests is still a common attribute. However, technology now allows the removal of more and more of this woody debris from the forest. As well, conversion of forests from old growth to managed stands shortens the rotation age from centuries to decades, thereby reducing the size and age of the trees left in the forest. This reduction drastically decreases the amount, size, and quality of dead and dying trees available for the future (Maser and Trappe 1984).

'Large, fallen trees are unique, critical components of forest systems' (Maser et al. 1979; Maser et al. 1988; Franklin and Hemstrom 1981; Franklin et al. 1981). However, in the early days of forest harvesting, dead wood was considered a hindrance to reforestation and stream access and quality (Triska and Cromack 1979). In the past, CWD was routinely removed in an attempt to limit fuel loading (thereby minimizing wildfires) and make replanting easier. Also, because of the slow rate at which CWD decayed, its role in nutrient cycling, and therefore its importance, was not well understood (Triska and Cromack 1979).

In the Pacific Northwest and British Columbia, removal of LOD in streams began as early as the mid-1800s and continued well into the mid-1900s. Large streams and rivers were used for navigation and transportation of logs to the mills, and were therefore kept clear of debris (Sedell et al. 1988). As logging encroached further up the stream and river valleys, so did LOD removal. When the streams became too small to transport logs, splash dams were built, allowing water to build up and sluice logs down to the larger rivers and streams (Bisson et al. 1987). Some rivers and streams were scoured down to bedrock, and have not yet recovered from the effects of these practices (Sedell et al. 1988).

In the 1950s and 1960s, fisheries managers were concerned that LOD restricted fish movement and was the cause of channel scouring during floods, besides creating logjams; they therefore prescribed its removal. The role of LOD in channel morphology was not yet understood (Sedell et al. 1988). Although some LOD removal is still prescribed to give fish access to the upper reaches of a stream, LOD is now considered an important part of a functioning fish stream.

Standing dead and dying trees (wildlife trees) were also routinely removed during timber harvest. They were once thought of only as fire and safety hazards that harboured insects and that were of no marketable value (Bull et al. 1986). Despite their newly recognized importance, standing dead or dying trees are still a threatened resource in British Columbia. They are still rapidly declining because of conventional silvicultural practices, fire prevention, firewood cutting, timber utilization standards, and worker safety regulations (Steeger and Machmer 1994).

Ecological Principles

Dead Wood Cycling
Dead wood cycling is the process of the cycling of components of wood (carbon, minerals, moisture, and so on) in the forest ecosystem through the processes of death, decomposition, and uptake. The changes to trees, which are the most significant structural features of forests, during this process affect many other forest components and functions (Maser et al. 1979; Maser and Trappe 1984; Maser et al. 1988; Hammond 1991). In the coastal ecosystems they studied, Franklin and Waring (1979) reported that approximately 17% of all ecosystem organic matter was found within logs (CWD) and standing dead trees (snags).

The dead wood cycle begins when the stem dies. Stem death may be immediate in some cases, such as lightning strikes, but this process is usually slow. The trees are usually the larger ones and remain standing for a relatively long period after they die. Recruitment of snags from the living tree population may vary with such factors as slope, aspect, rooting substrate, site moisture and nutrient conditions, tree species, and causes of mortality. These factors influence the longevity of snags, CWD, and LOD. The dynamics of each of these components are discussed below.

Dynamics of the Dead Wood Cycle

Snags

Thomas et al. (1979) and Backhouse (1993) describe decay classifications of snags and wildlife trees. Often the important characteristics of snags and wildlife trees are functions of previous disease or damage, and the result is a tree form or condition that is valuable to a variety of wildlife species.

In forests that are disturbed by processes other than forest harvesting, most trees eventually become dead wood; those that die and remain standing become part of the snag component of the forest. The causes of mortality in live trees are varied and are likely a combination of factors (Maser 1988). Mortality rates are usually species-specific (Franklin et al. 1987; Raphael and Morrison 1987) and are extremely variable because of multiple causal agents and varying site conditions. Mortality rates are reported to be greater in high-productivity sites than in low-productivity sites (Franklin et al. 1987). Often the mortality agents affecting trees are very different in different sites.

Researchers have reported regional differences in snag population dynamics. For example, stocking of dead standing trees in the Sub-Boreal Spruce (SBS) Zone (Meidinger and Pojar 1991) of interior British Columbia varied both regionally and locally. Mean stocking of standing dead trees greater than 7.5 cm dbh was 97.7 stems per hectare ($n = 51$) in the dry cool (dk) subzone and 218.8 ($n = 329$) in the moist cold (mc) subzone (Lofroth, unpublished data). Franklin et al. (1987) also reported regional differences in snag population dynamics. In mature stands within the SBSmc subzone, stocking of snags varied by as much as a factor of 25 among different site conditions. Dry ridgetop sites (Dry Pine) had the highest stocking of dead standing trees and alluvial sites (Cottonwood Bottomland) had the lowest. Analysis of size distributions revealed that much of this difference could be explained by mortality of small stems (< 20 cm). Furthermore, only sites that were mesic and wetter had dead standing stems greater than 40 cm dbh.

Although 'standing crop' may be important and useful in describing the structural characteristics of a site, rates of input and decay are just as important. Mortality rates are likely highest in younger seral stages (Franklin et al. 1987; Raphael and Morrison 1987; Lofroth, unpublished data), and primary causal agents likely change with succession (Franklin et al. 1987). Cline et al. (1980) reported that snag production rates and density (input) fell with increasing stand age, but mean snag size and longevity increased. Raphael and Morrison (1987) reported that decay rates for pines in their study area were greater than those for firs. They also reported that, regardless of species, falling rates declined with increasing tree diameter.

Dead standing trees are important components of the forest for species such as the Bald Eagle. (Photograph by Steve Voller)

In summary, recruitment of dead standing trees (snags and/or wildlife trees) varies depending on regional and local ecological conditions. Stem mortality (and therefore snag density) is often highest in younger seral stages. Snag densities also vary with ecological condition, but these relationships are influenced to a great extent by mortality agents. More productive sites generally have larger snags. Snag density decreases and longevity increases with increasing snag size.

Coarse Woody Debris (CWD)
Coarse woody debris enters the ecosystem either directly through the death and immediate fall of living trees (e.g., from windthrow), or through tree death and the eventual fall of standing dead material. As with standing dead trees, there is variability in the amount, size, species, and decay class of CWD. The amount of CWD in any stand is a function of mortality agents, site conditions and exposure, and decay mechanisms and rates (Harmon and Hua 1991). CWD biomass in some coniferous ecosystems may exceed the total biomass of many deciduous ecosystems (Spies and Cline 1988). Decay classification for CWD has been described by Maser et al. (1988).

The total volume or biomass of CWD varies with ecological condition (Spies and Cline 1988; Lofroth, unpublished data). Brewer (1993) reported that CWD in plots in some SBS stands ranged from 37 to 384 m³/ha. Mean volumes for all stand conditions and ages were 44.1 m³/ha in the SBSdk subzone and 159.2 m³/ha in the SBSmc subzone (Lofroth, unpublished data). The driest sites had the lowest biomass and moist sites the greatest in Douglas-fir forests of Washington and Oregon (Spies et al. 1988). Within mature stands in the SBSmc subzone, CWD volumes were lowest in xeric ecosystems (Dry Pine: 36.2 m³/ha) and highest in moist ecosystems (Devil's Club: 268.4 m³/ha) (Lofroth, unpublished data). Mean volumes of CWD in mature stands by natural disturbance type (Province of British Columbia 1995) and biogeoclimatic zone ranged from a low of 60 m³/ha in the Boreal Black and White Spruce Zone in *ecosystems with frequent stand-initiating events* (NDT3) (Province of British Columbia 1995) to a high of 390 m³/ha in the Coastal Western Hemlock Zone in *ecosystems with rare stand-initiating events* (NDT1) (Lofroth, unpublished data). Benson and Schlieter (1979) reported that CWD volumes were 210 m³/ha in dry-site Douglas-fir stands, but as much as 560 m³/ha in grand fir stands.

CWD volumes also vary with successional stage. Mean volumes across a range of moisture and nutrient regimes in the SBSmc subzone were high in early successional stages (herb/shrub) (174.2 m³/ha), declined to a low of 58.2 m³/ha in young forest successional stages, and were highest in old-growth stands (261.5 m³/ha) (Lofroth, unpublished data). Spies et al. (1988) reported that amounts of CWD were high in the youngest successional stages, were lowest in 60-80-year-old forests, and were high in old stands (< 500 years). After 500 years CWD amounts declined to an intermediate level. Spies and Franklin (1988) reported that CWD input may be low in young stands because of the small size of dead and dying stems. Volumes in these stands are often high, however, due to residual CWD from the previous stand. The amount of this residual CWD depends on the disturbance agent causing the change in succession. In stands where succession has been retarded by natural catastrophic events (windthrow, fire, etc.), it can be significant. Spies et al. (1988) suggest that the nature and the timing of disturbance play a key role in CWD dynamics. Human-caused changes (such as logging) will usually result in conditions different than those that may have initiated the original stand. In these circumstances, the amount of CWD, as with standing dead trees, may not be indicative of natural dynamics within ecosystems.

The rates of input and decay also vary with ecological site conditions and stand age, and between tree species (Sollins 1982; Spies and Franklin 1988). Sollins (1982) reported that although there was considerable variability in the data, the highest values for CWD biomass were reported from old-growth stands. Harmon et al. (1987) reported that decay rates of logs may vary with microclimate, size, substrate, and species of log. Mattson et al. (1987) reported that decay rates varied by as much as tenfold between tree species. They also reported that aspect was an important factor in determining decay rates, and that logs suspended above the ground decayed at slower rates than those on the ground. Keenan et al. (1993) attribute large accumulations of CWD in western redcedar and western hemlock stands on Vancouver Island to slow decomposition rates, high rates of input following windstorms, and the large size and decay resistance of western redcedar. Abbott and Crossley (1982) reported that in chestnut oak (*Quercus prinus*) stands, decomposition was influenced by moisture and temperature and was inversely related to the diameter of the material.

Large Organic Debris (LOD)
The dynamics of large organic debris (LOD) are influenced by factors that influence CWD dynamics and those related to hydrological features and processes. Maser and Trappe (1984) and Sedell et al. (1988) review LOD dynamics in detail. Thomson (1991) produced an annotated bibliography that reviewed a substantial portion of the literature on this topic. Much of this review is abridged from these accounts.

LOD enters aquatic ecosystems from a variety of sources, both chronic and episodic. Chronic input includes debris resulting from litterfall, individual tree mortality, and treefall; to some extent it is influenced by factors that are also important in CWD dynamics. Riparian ecosystems are the primary source of chronic input. Input from these mechanisms are similar in nature to CWD in streamside ecosystems, although some differences have been noted in abundance and piece size (Van Sickle and Gregory 1990). LOD also enters streams through episodic events such as floods, mass wasting, debris torrents, and large-scale mortalities due to events such as insect epidemics. Much of the material entering stream ecosystems is associated with the riparian zone, but some material originates in upslope forested ecosystems. McDade et al. (1990) measured source distances for logs entering small streams in a variety of stream classes and gradients in Oregon and Washington, and found that over 70% of the logs originated within 20 m of the stream channel. Chronic inputs of large material tend to be a feature of mature and old-growth ecosystems. LOD input from younger successional stages tends to be smaller in nature, but volumes may equal or exceed those originating from older stands. Input of LOD may also be influenced by the composition of vegetation. In the U.S. Pacific Northwest, red alder is a common early successional tree in riparian ecosystems and commonly becomes LOD. This is a consequence of its shallow roots and low resistance to undercutting (Bisson et al. 1987).

LOD 'output' from stream ecosystems is primarily a function of the physical breakdown of material and the transport of material downstream. Decay and breakdown processes in stream ecosystems are considerably slower than in terrestrial and marine ecosystems (Sedell et al. 1988). Biological decay processes are much less important in LOD dynamics than in CWD dynamics, primarily because of the anaerobic nature of the ecosystem. The retention time of LOD in stream ecosystems is influenced by the size and orientation of material, the nature of deposition, the size of the stream, the scale of major flooding events, the gradient of the stream, and the nature of sediment transfer in the stream (Maser and Trappe 1984; Bisson et al. 1987; Sedell et al. 1988; O'Connor and Ziemer 1989).

The Roles of Dead Wood in the Ecosystem

Dead wood (snags, CWD, LOD) functions in many ways in forested ecosystems. It provides habitat for a wealth of invertebrate, vertebrate, and plant species. It affects soil erosion, slope movement, pool and riffle formation, and nutrient capture and retention. Nutrient cycling in terrestrial and aquatic ecosystems is influenced by the characteristics of woody debris.

The following discussion of the relationship between dead wood and forest biota is strongly biased towards vertebrates. Information on plants and invertebrates is limited and incomplete. I briefly review the role of woody debris in a variety of ecosystem processes. The sources of information used to develop Tables 7.1 to 7.7 include Aubry et al. (1988), Backhouse and Lousier (1991), S. Berch (pers. comm., 1995), Campbell et al. (1990a, 1990b), S. Cannings (pers. comm., 1995), D.F. Fraser (pers. comm., 1995), Goward (1993), L. Friis (pers. comm., 1995), E.C. Lea (pers. comm., 1995), Lundquist and Mariani (1991), Nagorsen and Brigham (1993), J. Ptolemy (pers. comm., 1995), Redhead (1993), Ryan (1993), Ryan et al. (1993), Scudder (1994), Tripp (1994), and Terres (1991).

Ecological Roles of Snags

The relationship between cavity users (particularly birds) and snags may be one of the best documented wildlife/habitat relationships in North America (McLelland 1977; Thomas et al. 1979; Mannan et al. 1980; Davis et al. 1983; Mannan and Meslow 1984; Raphael and White 1984; Zarnowitz and Manuwal 1985; Lundquist and Mariani 1991; Machmer and Steeger 1993; and many others). Snags serve as nesting habitat for primary and secondary cavity-nesters (McLelland 1977; Thomas et al. 1979; and others) and perching habitat for many bird species. In British Columbia, 26 Red- and Blue-listed vertebrate species or subspecies depend on or are associated with snags for all or part of their life history (Tables 7.1 and 7.2). Characteristics that affect the value of individual snags as habitat include cause of death, diameter, tree form, bark condition, tree species, and height.

The cause of death affects the class and decay characteristics of snags. The ability of organisms such as fungi and insects to invade wildlife trees greatly affects the value of the snag for other wildlife. Decay organisms invading dead or dying trees serve to further weaken or soften the tree, allowing primary cavity-nesters to excavate nests (Thomas et al. 1979; Terres 1991).

The size of snags strongly influences which species may use them. Cavity-nesters in British Columbia include species as small as nuthatches, chickadees, and small bats to mammals as large as black bears (Tables 7.1 and 7.2) (Backhouse 1993). Black bears are known to use large cavities (particularly in western redcedar) as winter dens (Davis 1996). Fishers whelp almost exclusively in cavities in very large cottonwood trees in central B.C. (Weir 1995). Martens use cavities in trees as resting sites and maternal dens (Buskirk and Ruggiero 1994). Flying squirrels, red squirrels, and many other

Table 7.1

Mammals closely associated with snags in British Columbia

Species	Life history role	Management status
Western Long-eared Myotis (*Myotis evotis*)	roosting, maternal colonies	Yellow-listed
Northern Long-eared Myotis (*Myotis septentrionalis*)	roosting, maternal colonies	Red-listed
Keen's Long-eared Myotis (*Myotis keenii*)	roosting, maternal colonies	Red-listed
Western Red Bat (*Lasiurus blossevilli*)	roosting	Blue-listed
Hoary Bat (*Lasiurus cinereus*)	roosting	Yellow-listed
Silver-haired Bat (*Lasionycteris noctivagans*)	roosting, maternal colonies, hibernacula	Yellow-listed
Big Brown Bat (*Eptesicus fuscus*)	roosting, maternal colonies	Yellow-listed
Pallid Bat (*Antrozous pallidus*)	roosting	Red-listed
California Myotis (*Myotis californicus*)	roosting, maternal colonies	Yellow-listed
Western Small-footed Myotis (*Myotis ciliolabrum*)	roosting	Blue-listed
Little Brown Myotis (*Myotis lucifugus*)	roosting, maternal colonies	Yellow-listed
Long-legged Myotis (*Myotis volans*)	roosting, maternal colonies	Yellow-listed
Yuma Myotis (*Myotis yumanensis*)	roosting, maternal colonies	Yellow-listed
Southern Red-backed Vole (*Clethrionomys gapperi*)	nesting, summer dens	Yellow-listed

▶

◀ *Table 7.1*

Species	Life history role	Management status
Bushy-tailed Woodrat (*Neotoma cinerea*)	nesting, summer and winter dens	Yellow-listed
Deer Mouse (*Peromyscus maniculatus*)	nesting, summer and winter dens	Yellow-listed
Columbian Mouse (*Peromyscus oreas*)	nesting, summer and winter dens	Yellow-listed
Sitka Mouse (*Peromyscus sitkensis*)	nesting, summer and winter dens	Yellow-listed
Northern Flying Squirrel (*Glaucomys sabrinus*)	nesting, maternal and thermal dens	Yellow-listed
Yellow-pine Chipmunk (*Tamias amoenus*)	nesting, summer dens	Yellow-listed
Least Chipmunk (*Tamias minimus*)	nesting, summer dens	Yellow-listed; subsp. *selkirki* Red-listed; subsp. *oreocetes* Blue-listed
Red-tailed Chipmunk (*Tamias ruficaudus*)	nesting, feeding stations, maternal and thermal dens	Red-listed
Townsend's Chipmunk (*Tamias townsendii*)	nesting, summer dens	Yellow-listed
Douglas' Squirrel (*Tamiasciurus douglasii*)	nesting, feeding stations, maternal and thermal dens	Yellow-listed
Red Squirrel (*Tamiasciurus hudsonicus*)	nesting, feeding stations, maternal and thermal dens	Yellow-listed
Marten (*Martes americana*)	thermal and maternal dens	Yellow-listed
Fisher (*Martes pennanti*)	maternal dens	Blue-listed
Ermine (*Mustela erminea*)	maternal and summer dens	Yellow-listed
Long-tailed Weasel (*Mustela frenata*)	maternal and summer dens	Yellow-listed
Least Weasel (*Mustela nivalis*)	maternal and summer dens	Yellow-listed
Spotted Skunk (*Spilogale putorius*)	maternal, summer, and winter dens	Yellow-listed
Raccoon (*Procyon lotor*)	maternal, summer, and winter dens	Yellow-listed
Black Bear (*Ursus americanus*)	winter dens	Yellow-listed; subsp. *emmonsii* Blue-listed

Table 7.2

Birds closely associated with snags in British Columbia

Species	Life history role	Management status
Great Blue Heron (*Ardea herodias*)	nesting, roosting	Blue-listed
Wood Duck (*Aix sponsa*)	nesting	Yellow-listed
Common Goldeneye (*Bucephala clangula*)	nesting	Yellow-listed
Barrow's Goldeneye (*Bucephala islandica*)	nesting	Yellow-listed
Bufflehead (*Bucephala albeola*)	nesting	Yellow-listed
Hooded Merganser (*Lophodytes cucullatus*)	nesting	Yellow-listed
Common Merganser (*Mergus merganser*)	nesting	Yellow-listed
Turkey Vulture (*Cathartes aura*)	roosting	Blue-listed
Osprey (*Pandion haliaetus*)	nesting, roosting, perching	Yellow-listed
Bald Eagle (*Haliaeetus leucocephalus*)	nesting, roosting, perching	Blue-listed
Red-tailed Hawk (*Buteo jamaicensis*)	nesting, roosting, perching	Yellow-listed
Golden Eagle (*Aquila chrysaetos*)	nesting	Yellow-listed
American Kestrel (*Falco sparverius*)	nesting, perching	Yellow-listed
Merlin (*Falco columbarius*)	nesting	Yellow-listed
Barn Owl (*Tyto alba*)	nesting, winter roosting	Blue-listed
Flammulated Owl (*Otus flammeolus*)	nesting, roosting	Blue-listed
Western Screech Owl (*Otus kennicottii*)	nesting, roosting	Yellow-listed; subspp. *kennicottii* and *saturatus* Blue-listed
Great-horned Owl (*Bubo virginianus*)	nesting, perching	Yellow-listed
Northern Pygmy Owl (*Glaucidium gnoma*)	nesting	Yellow-listed; subsp. *swarthi* Blue-listed

▶

◀ *Table 7.2*

Species	Life history role	Management status
Northern Saw-Whet Owl (*Aegolius acadicus*)	nesting, roosting	Yellow-listed; subsp. *brooksi* Blue-listed
Spotted Owl (*Strix occidentalis*)	nesting, roosting	Red-listed
Barred Owl (*Strix varia*)	nesting, roosting	Yellow-listed
Boreal Owl (*Aegolius funereus*)	nesting, roosting	Yellow-listed
Vaux's Swift (*Chaetura vauxi*)	nesting, roosting	Yellow-listed
Belted Kingfisher (*Megaceryle alcyon*)	perching	Yellow-listed
Lewis Woodpecker (*Melanerpes lewis*)	nesting, roosting, perching, foraging	Blue-listed
Yellow-bellied Sapsucker (*Sphyrapicus varius*)	nesting, roosting, foraging	Yellow-listed
Red-naped Sapsucker (*Sphyrapicus nuchalis*)	nesting, roosting, foraging	Yellow-listed
Red-breasted Sapsucker (*Sphyrapicus ruber*)	nesting, roosting, foraging	Yellow-listed
Williamson's Sapsucker (*Sphyrapicus thyroideus*)	nesting, roosting, foraging	subsp. *thyroideus* Blue-listed; subsp. *nataliae* Red-listed
Downy Woodpecker (*Picoides pubescens*)	nesting, roosting, foraging	Yellow-listed
Hairy Woodpecker (*Picoides villosus*)	nesting, roosting, foraging	Yellow-listed; subsp. *picoideus* Blue-listed
White-headed Woodpecker (*Picoides albolarvatus*)	nesting, roosting, foraging	Red-listed
Three-toed Woodpecker (*Picoides tridactylus*)	nesting, roosting, foraging	Yellow-listed
Black-backed Woodpecker (*Picoides arcticus*)	nesting, roosting, foraging	Yellow-listed
Northern Flicker (*Colaptes auratus*)	nesting, roosting, foraging	Yellow-listed
Pileated Woodpecker (*Dryocopus pileatus*)	nesting, roosting, foraging	Yellow-listed
Pacific-slope Flycatcher (*Empidonax difficilis*)	nesting, perching	Yellow-listed

▶

◀ *Table 7.2*

Species	Life history role	Management status
Ash-throated Flycatcher (*Myriarchus cinerascens*)	nesting, perching	Yellow-listed
Purple Martin (*Progne subis*)	nesting, roosting	Red-listed
Tree Swallow (*Iridoprocne bicolor*)	nesting, roosting	Yellow-listed
Violet-Green Swallow (*Tachycineta thalassina*)	nesting, roosting	Yellow-listed
Black-capped Chickadee (*Parus atricapillus*)	nesting, roosting, foraging	Yellow-listed
Mountain Chickadee (*Parus gambeli*)	nesting, roosting, foraging	Yellow-listed
Boreal Chickadee (*Parus hudsonicus*)	nesting, roosting, foraging	Yellow-listed
Chestnut-backed Chickadee (*Parus rufescens*)	nesting, roosting, foraging	Yellow-listed
Red-breasted Nuthatch (*Sitta canadensis*)	nesting, roosting, foraging	Yellow-listed
White-breasted Nuthatch (*Sitta carolinensis*)	nesting, roosting, foraging	Yellow-listed
Pygmy Nuthatch (*Sitta pygmaea*)	nesting, roosting, foraging	Yellow-listed
Brown Creeper (*Certhia americana*)	nesting, roosting, foraging	Yellow-listed
Bewick's Wren (*Thryomanes bewickii*)	nesting	Yellow-listed
House Wren (*Troglodytes aedon*)	nesting	Yellow-listed
Western Bluebird (*Sialia mexicana*)	nesting, perching	Yellow-listed
Mountain Bluebird (*Sialia currucoides*)	nesting, perching	Yellow-listed
Common Grackle (*Quiscalus quiscula*)	nesting	Yellow-listed
House Finch (*Carpodacus mexicanus*)	nesting	Yellow-listed

mammalian species utilize cavities in snags and trees for part of their life history. Herpetofauna will use the space between loose bark and the trunk. Species such as Pileated Woodpeckers in western North America require trees in the larger diameter classes for the ecological conditions found there (Thomas et al. 1979; Bull et al. 1992; Bull and Holthausen 1993).

Snags are used by many species such as raccoons for cover and denning. (Courtesy B.C. Ministry of Environment, Lands and Parks)

Snags also provide perches for birds of prey and insectivores such as hawking flycatchers. Tree form and height are often important features of perches.

Bark retention and condition also influence the value of a snag as wildlife habitat. Species such as nuthatches, a variety of bats (including Red-listed Keen's long-eared myotis [*Myotis keenii*] and Northern long-eared myotis [*Myotis septentrionalis*]), and the clouded salamander use the space between sloughing bark and the tree bole as roosting and thermal habitat (Davis and Gregory 1993; Nagorsen and Brigham 1993). Some plant species, such as licorice fern (*Polypodium glycyrrhiza*), are epiphytic on snags. Others may be snag obligates (D.F. Fraser, pers. comm., 1995).

Different trees species differ in their value to wildlife. Harestad and Keisker (1989) reported a preference for aspen as a primary cavity-nesting tree in southern B.C. Lundquist and Mariani (1991) report that in the southern Washington Cascade Range, white pine snags were particularly important for woodpeckers and creepers, while Douglas-fir and western hemlock were more valuable for chickadees and nuthatches.

Ecological Roles of Coarse Woody Debris
Coarse woody debris plays numerous roles in providing habitat for organisms in forested ecosystems. Logs become habitat for a variety of invertebrate species shortly after falling. CWD is used by invertebrates as a source of food, for nesting and brooding sites, for protection from predators and environmental extremes, as a source of construction material, and as overwintering and hibernating sites (Samuelsson et al. 1994). Many invertebrates use or require particular species of CWD, and different communities of invertebrates occupy and use different decay stages of CWD (Harmon et al. 1986; Samuelsson et al. 1994). Insectivorous species such as woodpeckers, small mammals, and bears forage on insects dwelling in CWD (Maser et al. 1979; Maser and Trappe 1984; Samuelsson et al. 1994) (Tables 7.3 and 7.4).

Coarse woody debris provides thermal and security cover for a variety of small mammals in British Columbia. Sound CWD provides secure travel corridors for small mammals (Maser et al. 1979; Maser and Trappe 1984; Carter 1993), and provides subnivean habitat during winter. The value of this habitat is positively correlated with piece size (Maser and Trappe 1984; Hayes and Cross 1987; Carter 1993). Nordyke and Buskirk (1991) found that southern red-backed vole abundance was positively correlated with the decay stage of logs in the central Rocky Mountains. Maser and Trappe (1984) and Rhoades (1986) reported associations of small mammals with CWD because of the food source provided by the fungal fruiting bodies growing in and on the CWD.

Gyug (1993) reported that fur-bearers (martens and weasels) used clearcuts with logging debris more than those with no CWD; however, the level of use was much less than that of the adjacent forest. The value of CWD to mustelids (particularly martens, weasels, and fishers) is well documented (Baker 1992; Corn and Raphael 1992; Lofroth 1993; Buskirk and Powell 1994; Buskirk and Ruggiero 1994; and others). Martens select habitats partly on the basis of thermal microhabitats (Taylor 1993), such as those provided by CWD (Lofroth 1993; Buskirk and Powell 1994; Buskirk and Ruggiero 1994). Corn and Raphael (1992) reported that martens selected subnivean access points that had greater volumes of CWD, more layering of logs, more sound and moderately decayed logs, and fewer highly decayed logs than random sites.

Table 7.3

Mammals closely associated with coarse woody debris in British Columbia

Species	Life history role	Management status
Dusky Shrew (*Sorex obscurus*)	cover, foraging, reproduction	Yellow-listed
Water Shrew (*Sorex palustris*)	cover, foraging, reproduction	Yellow-listed
Vagrant Shrew (*Sorex vagrans*)	cover, foraging, reproduction	Yellow-listed
Shrew Mole (*Neurotrichus gibbsii*)	cover	Yellow-listed
California Myotis (*Myotis californicus*)	foraging	Yellow-listed
Snowshoe Hare (*Lepus americanus*)	cover	Yellow-listed
Southern Red-backed Vole (*Clethrionomys gapperi*)	cover, foraging	Yellow-listed; subsp. *occidentalis* Red-listed
Northern Red-backed Vole (*Clethrionomys rutilas*)	cover, foraging	Yellow-listed
Beaver (*Castor canadensis*)	dam construction	Yellow-listed
Cascades Golden-mantled Ground Squirrel (*Spermophilus saturatus*)	cover	Yellow-listed
Deer Mouse (*Peromyscus maniculatus*)	cover	Yellow-listed
Columbian Mouse (*Peromyseus oreas*)	cover	Yellow-listed
Sitka Mouse (*Promyscus sitkensis*)	cover	Yellow-listed
Douglas' Squirrel (*Tamiasciurus douglasii*)	cover	Yellow-listed
Marten (*Martes americana*)	denning, foraging	Yellow-listed
Fisher (*Martes pennanti*)	denning	Blue-listed
Ermine (*Mustela erminea*)	denning, foraging, cover	Yellow-listed; subsp. *haidarum* Red-listed; subsp. *anguinae* Blue-listed
Long-tailed Weasel (*Mustela frenata*)	denning, foraging, cover	Subsp. *altifrontalis* Red-listed; trapped
Least Weasel (*Mustela nivalis*)	denning, foraging, cover	Yellow-listed

Table 7.4

Birds closely associated with coarse woody debris in British Columbia

Species	Life history role	Management status
Northern Flicker (*Colaptes auratus*)	foraging	Yellow-listed
Pileated Woodpecker (*Dryocopus pileatus*)	foraging	Yellow-listed
Ruffed Grouse (*Bonasa umbellus*)	drumming	Yellow-listed

Aubry et al. (1988) found that some species of salamander were most abundant around CWD. Dupuis (1993) concluded that salamander populations in logged areas were limited by available moist microhabitats, primarily because of a lack of large logs in intermediate and advanced stages of decay. Salamanders use logs as reproduction sites, as foraging sites, and for cover, and also lay their eggs in them (Table 7.5) (Samuelsson et al. 1994).

Coarse woody debris functions as seed beds or nurse logs for some trees species and many species of bryophytes, fungi, and lichens, and some flow-

Fur-bearers such as marten select habitats partly on the basis of thermal microhabitats such as those provided by coarse woody debris. (Courtesy B.C. Ministry of Environment, Lands and Parks)

ering plants (Table 7.6) (Samuelsson et al. 1994; D.F. Fraser, pers. comm., 1995; E.C. Lea, pers. comm., 1995). CWD, and the associated epiphytic bryophytes, act as both nutrient and moisture buffers for the ecosystems (FEMAT 1993). This buffering allows the slow release of water and nutrients to surrounding plants. In mature and old-growth coastal forests, a large proportion of western hemlock and Sitka spruce seedlings germinate and grow on CWD substrates (Harmon and Franklin 1989; G. Davis, pers. comm., 1994). In the Crowsnest Forest, 40-70% of natural seedlings were rooted in decayed wood in old growth and 24% were rooted in decayed wood in cutblocks (S. Berch, pers. comm., 1995). CWD may be important to the establishment of vascular plants around wet sites such as ponds and bogs (D.F. Fraser, pers. comm., 1995). Red huckleberry (*Vaccinium parvifolium*) is likely an obligate CWD user (D.F. Fraser, pers. comm., 1995; E.C. Lea, pers. comm., 1995).

Other species are either associated with CWD or perhaps with the fungi that use CWD as their parasitic intermediate, such as the gnome plant (*Hypopitis congestum*), candystick (*Allotropa virgata*), and other ericaceous species. Ryan and Fraser (1993) reported that cryptogam species richness in coastal Douglas-fir forests was strongly influenced by available substrate. In forested sites, the presence of CWD and rock substrates resulted in substantial increases in species richness. The review of Samuelsson et al. (1994) of CWD states that distinct succession of bryophyte and lichen communities occurs as trees die, fall, and decay. In B.C., known decomposer macrofungi

Table 7.5

Herpetofauna closely associated with coarse woody debris in British Columbia

Species	Life history role	Management status
Pacific Giant Salamander (*Dicamtodon tenebrosus*)	reproduction	Red-listed
Clouded Salamander (*Aneides ferreus*)	cover, reproduction	Yellow-listed
Coeur d'Alene Salamander (*Plethodon idahoensis*)	cover, reproduction	Red-listed
Western Red-backed Salamander (*Plethodon vehiculum*)	cover	Yellow-listed
Western Skink (*Eumeces skiltonianus*)	cover	Yellow-listed
Rubber Boa (*Charina bottae*)	cover	Blue-listed
Sharp-tailed Snake (*Contia tenuis*)	cover	Red-listed

Table 7.6

Vascular plants closely associated with coarse woody debris in British Columbia

Species	Life history role	Management status
Western Hemlock (*Tsuga heterophylla*)	germination	Commercially harvested
Sitka Spruce (*Picea sitchensis*)	germination	Commercially harvested
Red Huckleberry (*Vaccinium parvifolium*)	germination, growth	–
Gnome Plant (*Hypopitis congestum*)	growth	–
Candystick (*Allotropa virgata*)	growth	Blue-listed

that are dependent on CWD include 162 species of bracket or shelf fungi/ conks, 364 species of other macrofungi, and some commercially harvested mushrooms, such as oyster mushrooms (S. Berch, pers. comm., 1995). These communities play roles in the germination and growth of other epiphytic and quasi-epiphytic communities. Climatic factors influence epiphytic communities, with lichens dominating drier ecosystems and bryophytes replacing them as conditions become wetter.

The longevity of individual pieces of CWD is critical to the persistence of many species with poor dispersal abilities. Dispersal in many species is from one log to the next, so logs close to each other are required. Samuelsson et al. (1994) note that large logs play a more important role than small logs in the ecology of bryophytes and lichens. Large logs last longer, have greater surface area, and have higher, steeper sides that prevent ground-dwelling species from invading. They may also be important in providing a relatively litter-free substrate for the establishment of some species of cryptogams (D.F. Fraser, pers. comm., 1995).

Ecological Roles of Large Organic Debris
The value of woody debris in providing habitat for anadromous and other game fish in aquatic ecosystems has been well documented (Thomson 1991). Large organic debris (LOD) increases aquatic habitat diversity by acting as a physical barrier to water, retaining or detaining sediment and controlling gravel movement (Miller 1987); helping to create and maintain ponds, back channels, and side pools (Bustard and Narver 1975a; Bisson et al. 1987; Sedell et al. 1988); increasing pool size, frequency, and stability; helping to form complex habitats such as riffles and plunge pools (Hamilton 1991);

Table 7.7

Fish closely associated with large organic debris in British Columbia

Species	Life history role	Management status
Coho Salmon (*Oncorhynchus kisutch*)	cover; facilitates deposition of spawning gravel	Yellow-listed
Chinook Salmon (*Oncorhynchus tshawytscha*)	cover; facilitates deposition of spawning gravel	Yellow-listed
Cutthroat Trout (*Oncorhynchus clarkii*)	cover, foraging; facilitates deposition of spawning gravel	Yellow-listed
Rainbow Trout (*Oncorhynchus mykiss*)	cover, foraging; facilitates deposition of spawning gravel	Yellow-listed
Brown Trout (*Salmo trutta*)	cover, foraging; facilitates deposition of spawning gravel	Yellow-listed
Bull Trout (*Salvelinus confluentus*)	cover, foraging; facilitates deposition of spawning gravel	Blue-listed
Brook Trout (*Salvelinus fontinalis*)	cover, foraging; facilitates deposition of spawning gravel	Yellow-listed
Dolly Varden (*Salvelinus malma*)	cover, foraging; facilitates deposition of spawning gravel	Yellow-listed
Mountain Whitefish (*Prosopium williamsoni*)	cover, foraging; facilitates deposition of spawning gravel	Yellow-listed
Arctic Grayling (*Thymallus arcticus*)	cover, foraging	Blue-listed; Williston stock Red-listed
Northern Pike (*Esox lucius*)	foraging, cover	Yellow-listed
Lake Chub (*Caiesius plumbeio*)	cover	Yellow-listed
Redside Shiner (*Richardsonius balteatus*)	cover	Yellow-listed
Burbot (*Lota lota*)	cover	Yellow-listed
Brook Stickleback (*Culaea inconstans*)	cover	Yellow-listed
Threespine Stickleback (*Gasterosteus aculeatus*)	cover	Yellow-listed
Ninespine Stickleback (*Pungitius pungitius*)	cover	Yellow-listed
Smallmouth Bass (*Micropterus dolomieui*)	foraging, cover	Yellow-listed
Largemouth Bass (*Micropterus salmoides*)	foraging, cover	Yellow-listed

Table 7.8

_____

Mammals closely associated with large organic debris in British Columbia

Species	Life history role	Management status
Beaver (*Castor canadensis*)	dam construction	Yellow-listed
River Otter (*Lutra canadensis*)	cover, foraging	Yellow-listed
Mink (*Mustela vison*)	cover, foraging	Yellow-listed

_____

and providing a substrate for biological activity (Sedell et al. 1988). LOD helps regulate local water flow and depth, and increases water depth variability, providing preferred habitat for some species (Bustard and Narver 1975b).

LOD provides important cover for fish for hiding and resting (Hamilton 1991). Table 7.7 lists fish species in B.C. that are closely associated with LOD for all or part of their life history. LOD also acts as a food source and habitat for many types of aquatic invertebrates (Harmon et al. 1986). Dudley and Anderson (1982) documented 56 taxa of invertebrates closely associated with wood, and an additional 129 taxa that were dead-wood facultatives in stream ecosystems in the U.S. Pacific Northwest. Cummins and Klug (1979) and Maser and Trappe (1984) identified 5 functional groups of aquatic invertebrates reliant on LOD: borers/tunnellers; wood ingesters and shredders; algae scrapers; those that attach to wood or hide in its grooves; and piercers and predators. LOD provides substrate for algae and microbes, which in turn provide food for aquatic invertebrates (Maser and Trappe 1984). The state of decay of LOD is a critical factor in determining the biotic community that may take advantage of it as growing substrate, burrowing substrate, or food source. Aquatic invertebrates are a major food source for aquatic vertebrates. LOD also provides habitat for mammals, particularly beavers, mink, and otter (Table 7.8), and long-toed salamanders. LOD is used by some birds (Kingfisher, American Dipper, Wood Duck, Hooded Merganser), but is likely less critical to them.

Pools formed by LOD act as collection basins for finer organic matter. The size and position of debris is correlated with the size and amount of pool habitat formed (Sedell et al. 1988). Trapped organic matter, such as leaves and needles, forms much of the energy input into stream ecosystems and may be the dominant regulator of ecosystem organic 'output' (Bilby and Likens 1980; Bilby 1981, 1984). Stream rehabilitation after major floods, debris torrents, or massive landslides is accelerated by large, woody debris along and within the channel (Sedell et al. 1988). Plant species diversity on river bars is related to the area, sediment, and woody debris of river bars (Malanson and Butler 1990).

Dead Wood and Ecosystem Processes

Dead wood is a critical component of many ecosystem processes. It supports physical, chemical, and biological functions in forested ecosystems. These functions include nutrient cycling, carbon storage, erosion control and slope stabilization, water cycling, soil formation, and stream movement processes (Harmon et al. 1986; Maser et al. 1988; Caza 1993; Samuelsson et al. 1994).

Coarse woody debris is a significant factor in nutrient cycling processes (Harmon et al. 1986; Caza 1993). Although the relative concentration of nutrients in wood and bark is low, much of the nutrient capital and carbon are stored here because of the large biomass involved (Harmon et al. 1986; Caza 1993). Dead wood facilitates a slow release of nutrients, ameliorates leaching, and provides a growing substrate for bryophytes. These buffer water and nutrient release from litterfall and above-ground processes, especially processes such as nitrogen fixation in above-ground plants such as hepatics (Harmon et al. 1986; FEMAT 1993; Samuelsson et al. 1994). Free-living bacteria in woody residues and soil wood fix 30-60% of the nitrogen in the forest soil. In addition, 20% of soil nitrogen is stored in these components (Harvey et al. 1987). Harmon et al. (1986) reported that CWD accounted for as much as 45% of above-ground stores of organic matter.

Dead wood in terrestrial ecosystems is a primary location for fungal colonization and often acts as refugia for mycorrhizal fungi during ecosystem disturbance (Triska and Cromack 1979; Harmon et al. 1986; Caza 1993). Colonization of dead wood by fungi and microbes may be one of the most important stages in nutrient cycling (Caza 1993); however, these processes are still relatively poorly understood. Soil wood contains a disproportionate amount of the coniferous feeder roots or ectomycorrhizae in forests (Harvey et al. 1987). As one of the dominant sources of organic matter, dead wood is an important determinant in soil formation and composition (Caza 1993).

Dead wood is also the dominant store of organic matter in stream ecosystems (Harmon et al. 1986); as such, it is an important source of nutrient and organic matter input. Dead wood traps leaf and litterfall within aquatic systems, which extends the length of time this material remains and provides nutrients through decomposition (Triska and Cromack 1979; Harmon et al. 1986).

Dead wood provides physical structure to the ecosystem and fills such roles as sediment storage (Wilford 1984), protecting the forest floor from mineral soil erosion and mechanical disturbance during harvesting activities. It ameliorates the effects of cold air drainage on plants, helps stabilize slopes, and minimizes soil erosion (Maser et al. 1988). Dead wood provides elevated germination platforms with reduced litterfall accumulation and relatively consistent moisture regimes (Harmon et al. 1986; Maser et al. 1988; Caza 1993; D.F. Fraser, pers. comm., 1995). In stream ecosystems it protects stream banks from erosion and maintains channel stability (Triska and Cromack 1979; Sedell et al. 1988). Features that influence the ability of LOD to fulfil these functions include size (length and diameter), whether roots are still attached, orientation, degree of burial, and proportion of the piece that remains submerged (Sedell et al. 1988).

In both terrestrial and aquatic ecosystems, dead wood functions as a reservoir of moisture, ameliorating drought conditions and providing a 'perched water table' (Triska and Cromack 1979).

Management of the Dead Wood Cycle

Almost all aspects of forest management affect the dead wood cycle. Harvesting initially increases CWD amounts except where utilization is very high. CWD may be harvested where utilization standards influence the amounts of waste and avoidable waste retained in second-growth forests. Site preparation (such as broadcast burning, windrowing, or piling and burning) leave structurally simpler forest floor ecosystems. These ecosystems lack or have significantly reduced amounts of CWD, and recruitment of new CWD from the regenerating stand may not occur until after the projected rotation age is reached.

Public perception of CWD as messy logging that wastes wood has influenced CWD management. This has led to a policy of 'zero waste tolerance.' The importance of LOD in stream ecosystems and the role of snags are more widely accepted. Management of CWD requires increased understanding of its importance in the forest management arena, the environmental community, and the general public.

Management of the dead wood cycle must ensure continued input of material into forested ecosystems (terrestrial and aquatic) so that the important biological, chemical, and physical functions are fulfilled. The previous sections point out the need to address a variety of attributes or characteristics of dead wood to achieve this goal.

The value of snags, CWD, and LOD as a supply of habitat and as components of ecosystem processes is correlated to a large extent with the amount of material and the piece size. As with most attributes, however, there is a succession of biota and ecosystem functions with increasing piece size. Decay characteristics are another important consideration in managing the dead wood cycle. Distinct biotic communities are associated with different decay states, and the role of dead wood in ecosystem function changes considerably as decay progresses (i.e., from sediment storage and aquatic pool formation to nutrient cycling). Species composition of dead wood is another attribute that must be addressed.

Use by biota, input and decay processes, and nutrient and moisture storage functions may all be species-dependent. Orientation, distribution, and structural arrangement in the terrestrial and aquatic landscape influences the value of dead wood in the ecosystem, particularly in processes such as slope stabilization, sediment control, stream bank stabilization, fungal recolonization, and habitat value for some species.

Management of the dead wood cycle requires an ecosystem approach because of the dead wood's ecological variability. Management regimes need to be designed to ensure the persistence of the full range of natural variation for each of the following characteristics of dead wood: amount, size, species, decay class, orientation, distribution, and structural arrangement. Currently little information is available for effective management. The next section outlines the types of information still needed.

Because wildlife trees and snags are an important source of input to the dead wood cycle, appropriate management of these components is critical. A system is in place in British Columbia to guide the management of wildlife trees and snags at the landscape and stand scales (Province of British Columbia 1995). There is a sanctioned (B.C. Ministry of Forests and B.C. Workers' Compensation Board) process for assessing and maintaining snags in silviculture operations with low ground vibration, such as tree planting and juvenile spacing, and in certain harvesting scenarios (Guy and Manning 1995). Currently there are no ecosystem management systems or prescriptions in place for the management of CWD and LOD. Management regimes must be designed to ensure that dead wood input and decay processes satisfy the criteria for the factors discussed above, and are explicitly addressed by harvesting and silviculture plans and operations.

Research Needs

Information needs for ecosystem-based management would best be met by a combination of research and inventory tasks that address the following topics:

Dynamics of wildlife trees and snags: natural mortality rates of trees by species and size class, and the development of life tables for snags to assess longevity in the ecosystem.

Dynamics of CWD: rates of input, decay, size and species distribution, orientation, and distribution in natural and managed stands.

Dynamics of LOD, particularly input characteristics of size, species, orientation, and relative rates of loss through decay and downstream transport.

Development of models for dead wood cycle dynamics that could be integrated with tree growth and timber supply models.

The nature of relationships between dead wood and dead wood-obligate species, particularly taxa that are still poorly understood and undescribed, such as bryophytes, hepatics, lichens, and invertebrates.

Importance of dead wood in ecosystem processes in different ecological conditions in B.C.

Pre- and post-assessments of the nature of dead wood during forest-harvesting operations in a variety of ecological conditions. This information will highlight attributes and ecological areas that need immediate management attention.

Examination of the effect of a range of silvicultural treatments on the dynamics and ecosystem value of dead wood.

Relationships between CWD dynamics and Ministry of Forests waste and residue assessment procedures in order to clarify the potential overlap between these.

Assessment of the extent to which wildlife tree and snag management will ensure an adequate input of dead wood to the forested ecosystem.

Literature Cited

Abbott, D.T., and D.A. Crossley Jr. 1982. Woody litter decomposition following clear-cutting. Ecology 63(1):35-42.

Aubry, K.B., L.L.C. Jones, and P.A. Hall. 1988. Use of woody debris by plethodontid salamanders in Douglas-fir forests in Washington. Pages 32-37 in R.C. Szaro, K.E. Severson, and D.R. Patton (tech. coords.). Management of amphibians, reptiles and small mammals in North America. USDA Forest Service General Technical Report RM-166. Fort Collins, Colo.

Backhouse, F. 1993. Wildlife tree management in British Columbia. Co-publication of B.C. Environment, Workers' Compensation Board, B.C. Forest Service and FRDA. Victoria, B.C. 32 pp.

Backhouse, F., and J.D. Lousier. 1991. Silviculture systems research: wildlife tree problem analysis. B.C. Ministry of Forests, B.C. Ministry of Environment, and B.C. Wildlife Tree Committee, Victoria, B.C. 205 pp.

Baker, J.M. 1992. Habitat use and spatial organization of pine marten on southern Vancouver Island, British Columbia. M.Sc. thesis, Simon Fraser University, Burnaby, B.C. 119 pp.

Benson, R.E., and J.A. Schlieter. 1979. Woody material in northern Rocky Mountain forests: volume, characteristics, and changes with harvesting. Pages 27-36 in Environmental consequences of timber harvesting in rocky mountain coniferous forests. Symposium proceedings. 11-13 September 1979, Missoula, Mont. USDA Forest Service General Technical Report INT-90. Ogden, Utah. 526 pp.

Bilby, R.E. 1981. Role of organic debris dams in regulating the export of dissolved and particulate matter from a forested watershed. Ecology 62(5):1234-1243.

–. 1984. Removal of woody debris may affect stream channel stability. Journal of Forestry 82(10):609-613.

Bilby, R.E., and G.E. Likens. 1980. Importance of organic debris dams in the structure and function of stream ecosystems. Ecology 61(5):1107-1113.

Bisson, P.A., R.E. Bilby, M.D. Bryand, C.A. Dolloff, G.B. Grette, R.A. House, M.L. Murphy, K.V. Koski, and J.R. Sedell. 1987. Large woody debris in forested streams in the Pacific

Northwest: past, present and future. Pages 143-189 in E.O. Salo and T.W. Cundy (eds.). Streamside management: forestry and fishery interactions. University of Washington Institute of Forestry Resources, Contribution No. 57. Seattle.

Brewer, R.G.W. 1993. Characterization and orientation of coarse woody debris in some old Sub-Boreal Spruce stands. B.Sc. thesis, University of British Columbia, Vancouver, B.C. 91 pp.

Bull, E.L., and R.S. Holthausen. 1993. Habitat use and management of Pileated Woodpeckers in northeastern Oregon. Journal of Wildlife Management 57(2):335-345.

Bull, E.L., R.S. Holthausen, and M.G. Henjum. 1992. Roost trees used by Pileated Woodpeckers in northeastern Oregon. Journal of Wildlife Management 56(4):786-793.

Buskirk, S.W., and R.A. Powell. 1994. Habitat ecology of fishers and American martens. Pages 283-296 in S.W. Buskirk, A.S. Harestad, and M.G. Raphael (comps. and eds.). Martens, sables, and fishers: biology and conservation. Cornell University Press, Ithaca, N.Y. 484 pp.

Buskirk, S.W., and L.F. Ruggiero. 1994. American Marten. Pages 7-37 in L.F. Ruggiero, K.B. Aubry, S.W. Buskirk, L.J. Lyon, and W.J. Zielinski (tech. eds.). The scientific basis for conserving forest carnivores. American marten, fisher, lynx, and wolverine in the western United States. USDA Forest Service General Technical Report RM-254. Fort Collins, Colo. 184 pp.

Bustard, D.R., and D.W. Narver. 1975a. Aspects of the winter ecology of juvenile coho salmon (*Oncorhynchus kisutch*) and steelhead trout (*Salmo gardneri*). Journal of the Fisheries Research Board of Canada 32:667-680.

–. 1975b. Preferences of juvenile coho salmon (*Oncorhynchus kisutch*) and cutthroat trout (*Salmo clarki*) relative to simulated alteration of winter habitat. Journal of the Fisheries Research Board of Canada 32:681-687.

Campbell, R.W., N.K. Dawe, I. McTaggart-Cowan, J.M. Cooper, G.W. Kaiser, and M.C.E. McNall. 1990a. The birds of British Columbia. Vol. 1: introduction, loons through waterfowl. Canadian Wildlife Service and Royal British Columbia Museum. Victoria, B.C. 514 pp.

–. 1990b. The birds of British Columbia. Vol. 2: nonpasserines: diurnal birds of prey through woodpeckers. Canadian Wildlife Service and Royal British Columbia Museum. Victoria, B.C. 636 pp.

Carter, D.W. 1993. The importance of seral stage and coarse woody debris to the abundance and distribution of deer mice on Vancouver Island, British Columbia. M.Sc. thesis, Simon Fraser University, Burnaby, B.C. 105 pp.

Caza, C.L. 1993. Woody debris in the forests of British Columbia: a review of the literature and current research. B.C. Ministry of Forests Land Management Report No. 78. Victoria, B.C. 99 pp.

Cline, S.P., A.B. Berg, and H.M. Wight. 1980. Snag characteristics and dynamics in Douglas-fir forests, western Oregon. Journal of Wildlife Management 44(4):773-786.

Corn, J.G., and M.G. Raphael. 1992. Habitat characteristics at marten subnivean access sites. Journal of Wildlife Management 56(3):442-448.

Cummins, K.W., and M.J. Klug. 1979. Feeding ecology of stream invertebrates. Annual Review of Ecological Systems 10:147-172.

Davis, H. 1996. Characteristics and selection of winter dens by black bears in coastal British Columbia. M.Sc. thesis, Simon Fraser University, Burnaby, B.C. 147 pp.

Davis, T.M., and P.T. Gregory. 1993. Status of the clouded salamander in British Columbia. B.C. Environment Wildlife Working Report No. WR-53. Victoria, B.C. 13 pp.

Davis, J.W., G.A. Goodwin, and R.A. Ockenfels (tech. coords.). 1983. Snag habitat management: proceedings of the symposium. USDA Forest Service General Technical Report RM-99. 226 pp.

Dudley, T., and N.H. Anderson. 1982. A survey of invertebrates associated with wood debris in aquatic habitats. Technical Paper 6419. Oregon Agricultural Experiment Station, Corvallis, Ore.

Dupuis, L.A. 1993. The status and distribution of terrestrial amphibians in old-growth forests and managed stands. M.Sc. thesis, University of British Columbia, Vancouver, B.C. 66 pp.

Forest Ecosystem Management Assessment Team (FEMAT). 1993. Forest ecosystem management: an ecological, economic and social assessment. USDA Forest Service (and other agencies), Portland, Ore.

Franklin, J.F., H.H. Shugart, and M.E. Harmon. 1987. Tree death as an ecological process. BioScience 37(8):550-556.

Franklin, J.F., and R.H. Waring. 1979. Distinctive features of the northwestern coniferous forest: development, structure and function. Pages 59-86 in R. Waring (ed.). Forests: fresh perspectives from ecosystem analysis. Proceedings of the 40th Annual Biology Colloquium. Oregon State University Press, Corvallis, Ore.

Goward, T. 1993. Lichen inventory requirements. Part 2 of M. Ryan, T. Goward, and S. Redhead. Nonvascular plant inventory requirements for British Columbia. Unpublished manuscript. Arenaria Research and Interpretation, Victoria, B.C.

Guy, S.E., and E.T. Manning. 1995. Wildlife/danger tree assessor's course workbook. 4th ed. B.C. Ministry of Forests. Victoria, B.C.

Gyug, L.W. 1993. Furbearer and prey use of logging debris piles in clearcuts: progress report 1992/93. Unpublished manuscript. B.C. Ministry of Environment, Lands and Parks. 25 pp.

Hamilton, S.R. 1991. Streamside management zones and fisheries protection. Pages 45-49 in Proceedings WildFor91. Wildlife and forestry: towards a working partnership. Canadian Society of Environmental Biologists and Canadian Pulp and Paper Association.

Hammond, H. 1991. Seeing the forest among the trees: the case for wholistic forest use. Polestar Press, Vancouver, B.C. 309 pp.

Harestad, A.S., and D.G. Keisker. 1989. Nest tree use by primary cavity-nesting birds in south central British Columbia. Canadian Journal of Zoology 67:1067-1073.

Harmon, M.E., and J.F. Franklin. 1989. Tree seedlings on logs in *Picea-Tsuga* forests of Oregon and Washington. Ecology 70(1):48-59.

Harmon, M.E., K. Cromack Jr., and B.G. Smith. 1987. Coarse woody debris in mixed-conifer forests, Sequoia National Park, California. Canadian Journal of Forestry Research 17:1265-1272.

Harmon, M.E., J.F. Franklin, F.J. Swanson, P. Sollins, S.V. Gregory, D. Lattin, N.H. Anderson, S.P. Cline, N.G. Aumen, J.R. Sedell, W. Lienkaemper, K. Cromack Jr., and K.W. Cummings. 1986. Ecology of coarse woody debris in temperate ecosystems. Advances in Ecological Research 5:133-302.

Harmon, M.E., and C. Hua. 1991. Coarse woody debris dynamics in two old-growth ecosystems. BioScience 41(9):604-610.

Harvey, A.E., M.F. Jurgensen, M.J. Larsen, and R.T. Graham. 1987. Decaying organic materials and soil quality in the inland northwest: a management opportunity. USDA Forest Service General Technical Report INT-225.

Hayes, J.P., and S.P. Cross. 1987. Characteristics of logs used by western red-backed voles, *Clethrionomys californicus*, and deer mice, *Peromyscus maniculatus*. Canadian Field Naturalist 101(4):543-546.

Keenan, R.J., and A. Inselberg. (in prep.). Coarse woody debris in a forested watershed on the west coast of Vancouver Island, British Columbia.

Keenan, R.J., C.E. Prescott, and J.P. Kimmins. 1993. Mass and nutrient content of woody debris and forest floor in western redcedar and western hemlock forests on northern Vancouver Island. Canadian Journal of Forestry Research 23:1052-1059.

Lofroth, E.C. 1993. Scale dependent analyses of habitat selection by marten in the Sub-Boreal Spruce biogeoclimatic zone, British Columbia. M.Sc. thesis, Simon Fraser University, Burnaby, B.C. 109 pp.

Lundquist, R.W., and J.M. Mariani. 1991. Nesting habitat and abundance of snag-dependent birds in the southern Washington Cascade Range. Pages 221-241 in L.F. Ruggiero, K.B. Aubry, A.B. Carey, and M.H. Huff (eds.). Wildlife and vegetation of unmanaged Douglas-fir forests. USDA Forest Service General Technical Report PNW-GTR-285. Portland, Ore.

Machmer, M.M., and C. Steeger. 1993. The ecological roles of wildlife tree users in forest ecosystems. Unpublished manuscript prepared for B.C. Ministry of Forests, Prince George, B.C. 72 pp.

Malanson, G.P., and D.R. Butler. 1990. Woody debris, sediment and riparian vegetation of a subalpine river, Montana, USA. Arctic and Alpine Research 22(2):183-194.

Mannan, R.W., and E.C. Meslow. 1984. Bird populations and vegetation characteristics in managed and old-growth forests, northeastern Oregon. Journal of Wildlife Management 48(4):1219-1238.

Mannan, R.W., E.C. Meslow, and H.M. Wight. 1980. Use of snags by birds in Douglas-fir forests, western Oregon. Journal of Wildlife Management 44(4):787-797.

Maser, C. 1988. The redesigned forest. R&E Miles, San Pedro, Calif. 234 pp.

Maser, C., and J.M. Trappe. 1984. The seen and unseen world of the fallen tree. USDA Forest Service General Technical Report PNW-164. 56 pp.

Maser, C., R.G. Anderson, K. Cromack Jr., J.T. Williams, and R.E. Martin. 1979. Dead and down material. Pages 78-95 in J.W. Thomas (tech. ed.). Wildlife habitats in managed forests: the Blue Mountains of Oregon and Washington. USDA Agricultural Handbook No. 553. 512 pp.

Maser, C., S.P. Cline, K. Cromack Jr., J.M. Trappe, and E. Hansen. 1988. What we know about large trees that fall to the forest floor. Pages 25-46 in C. Maser, R.F. Tarrant, J.M. Trappe, and J.F. Franklin (eds.). From the forest to the sea: a story of fallen trees. USDA Forest Service General Technical Report PNW-GTR-229. Published in cooperation with the USDI Bureau of Land Management. Portland, Ore. 153 pp.

Mattson, K.G., W.T. Swank, and J.B. Waide. 1987. Decomposition of woody debris in a regenerating, clear-cut forest in the Southern Appalachians. Canadian Journal of Forest Research 17:72-721.

McClelland, B.R. 1977. Relationships between hole-nesting birds, forest snags, and decay in western larch–Douglas-fir forests of the northern Rocky Mountains. Ph.D. thesis, University of Montana, Missoula, Mont. 438 pp.

McDade, M.H., F.J. Swanson, W.A. McKee, J.F. Franklin, and J. Van Sickle. 1990. Source distances for coarse woody debris entering small streams in western Oregon and Washington. Canadian Journal of Forest Research 20:326-330.

McRae, D.J., M.E. Alexander, and B.J. Stocks. 1979. Measurement and description of fuels and fire behaviour on prescribed burns: a handbook. Canadian Forestry Service, Great Lakes Forest Research Centre Report 0-X-287.

Meidinger, D., and J. Pojar (eds.). 1991. Ecosystems of British Columbia. B.C. Ministry of Forests Special Report Series No. 6. 330 pp.

Miller, E. 1987. Effects of forest practices on relationships between riparian area and aquatic ecosystems. Pages 40-47 in Managing southern forests for wildlife and fish: a proceedings. January 1987, New Orleans, La. USDA Forest Service General Technical Report SO-65.

Nagorsen, D.W., and R.M. Brigham. 1993. The bats of British Columbia. Royal British Columbia Museum, Victoria, B.C., and UBC Press, Vancouver, B.C. 165 pp.

Nordyke, K.A., and S.W. Buskirk. 1991. Southern red-backed vole, *Clethrionomys gapperi*, populations in relation to stand succession and old-growth character in the central Rocky Mountains. Canadian Field Naturalist 105(3):330-334.

O'Connor, M.D., and R.R. Ziemer. 1989. Coarse woody debris ecology in a second-growth *Sequoia sempervirens* forest stream. Pages 165-171 in Proceedings of the California Riparian Systems Conference. 22-24 September 1988, Davis, Calif. USDA Forest Service General Technical Report PSW-110.

Province of British Columbia. 1995. Forest Practices Code of British Columbia Biodiversity Guide Book. Ministry of Forests and B.C. Environment. Victoria, B.C. 99 pp.

Quesnel, H. 1994. Assessment and characterization of old-growth stands in the Nelson Forest Region – progress report. B.C. Forest Service, Nelson, B.C.

Raphael, M.G., and M.L. Morrison. 1987. Decay and dynamics of snags in the Sierra Nevada, California. Forest Science 33(3):774-783.

Raphael, M.G., and M. White. 1984. Use of snags by cavity-nesting birds in the Sierra Nevada. Wildlife Monographs No. 86. 66 pp.

Redhead, S. 1993. Macrofungi inventory requirements. Part 3 of M. Ryan, T. Goward, and S. Redhead. Nonvascular plant inventory requirements for British Columbia. Unpublished manuscript. Arenaria Research and Interpretation, Victoria, B.C.

Rhoades, F. 1986. Small mammal mycophagy near woody debris accumulations in the Stehekin River valley, Washington. Northwest Science 60(3):150-153.

Ryan, M.W. 1993. Bryophyte inventory requirements. Part 1 of M. Ryan, T. Goward, and S. Redhead. Nonvascular plant inventory requirements for British Columbia. Unpublished manuscript. Arenaria Research and Interpretation, Victoria, B.C.

Ryan, M.W., and D.F. Fraser. 1993. Changes in plant diversity in Douglas-fir stands following the conversion of old growth to second growth. Pages 16-20 in V. Marshall (ed.). Proceedings of the Forest Ecosystem Dynamics Workshop. B.C. Ministry of Forests and Forestry Canada. FRDA 210. 98 pp.

Ryan, M.W., T. Goward, and S. Redhead. 1993. Nonvascular plant inventory requirements for British Columbia. Unpublished manuscript. Arenaria Research and Interpretation. Victoria, B.C.

Samuelsson, J., L. Gustafsson, and T. Ingelog. 1994. Dying and dead trees: a review of their importance for biodiversity. Swedish Threatened Species Unit, Uppsala, Sweden. 110 pp.

Scudder, G.G.E. 1994. Priorities for inventory and descriptive research on British Columbia terrestrial and freshwater invertebrates. Unpublished manuscript. 191 pp.

Sedell, J.R., P.A. Bisson, F.J. Swanson, and S.V. Gregory. 1988. What we know about large trees that fall into streams and rivers. Pages 47-81 in C. Maser, R.F. Tarrant, J.M. Trappe, and J.F. Franklin (eds.). From the forest to the sea: a story of fallen trees. USDA Forest Service General Technical Report PNW-GTR-229. Published in cooperation with the USDI Bureau of Land Management, Portland, Ore. 153 pp.

Sollins, P. 1982. Input and decay of coarse woody debris in coniferous stands in western Oregon and Washington. Canadian Journal of Forest Research 12:18-28.

Spies, T.A., and S.P. Cline. 1988. Coarse debris in forests and plantations of coastal Oregon. Pages 5-10 in C. Maser, R.F. Tarrant, J.M. Trappe, and J.F. Franklin (eds.). From the forest to the sea: a story of fallen trees. USDA Forest Service General Technical Report PNW-GTR-229. Published in cooperation with the USDI Bureau of Land Management, Portland, Ore. 153 pp.

Spies, T.A., and J.F. Franklin. 1988. Old growth and forest dynamics in the Douglas-fir region of western Oregon and Washington. Natural Areas Journal 8(3):190-201.

Spies, T.A., J.F. Franklin, and T.B. Thomas. 1988. Coarse woody debris in Douglas-fir forests of western Oregon and Washington. Ecology 69(6):1689-1702.

Steeper, C., and M.M. Machner. 1994. Inventory and description of wildlife trees and primary cavity nesters in selected stands of the Nelson Forest Region. B.C. Ministry of Forests, Nelson, B.C.

Steventon, J.D. 1994. Biodiversity of the Prince Rupert Forest Region and biodiversity and forest management in the Prince Rupert Forest Region: a discussion paper. B.C. Ministry of Forests Land Management Report No. 82. Victoria, B.C. 30 pp.

Terres, J.K. 1991. The Audubon Society encyclopedia of North American birds. Wing Books, New York. 1,109 pp.

Thomas, J.W., R.G. Anderson, C. Maser, and E.L. Bull. 1979. Snags. Pages 60-77 in J.W. Thomas (tech. ed.). Wildlife habitats in managed forests: the Blue Mountains of Oregon and Washington. USDA Agricultural Handbook No. 553. 512 pp.

Thomson, B. 1991. Annotated bibliography of large organic debris (LOD) with regards to stream channels and fish habitat. B.C. Ministry of Environment Technical Report 32. Victoria, B.C. 93 pp.

Tripp, T. 1994. The decay classes of coarse woody debris and the wildlife that use them. Unpublished manuscript prepared for B.C. Ministry of Forests, Integrated Resources Section. 14 pp.

Triska, F.J., and K. Cromack Jr. 1979. The role of wood debris in forests and streams. Pages 171-190 in R. Waring (ed.). Forests: fresh perspectives from ecosystem analysis. Proceedings of the 40th Annual Biology Colloquium. Oregon State University Press, Corvallis, Ore.

Trowbridge, R., B. Hawkes, A. Macadam, and J. Parminter. 1987. Field handbook for prescribed fire assessments in British Columbia: logging slash fuels. B.C. Ministry of Forests FRDA Handbook 001. 63 pp.

Van Sickle, J., and S.V. Gregory. 1990. Modeling inputs of large woody debris to streams from falling trees. Canadian Journal of Forestry Research 20:1593-1601.

Weir, R.D. 1995. Diet, spatial organization, and habitat relationships of fishers in south-central British Columbia. M.Sc. thesis, Simon Fraser University, Burnaby, B.C. 139 pp.

Wilford, D.J. 1984. The sediment-storage function of large organic debris at the base of unstable slopes. Pages 115-119 in W.R. Meehan, T.R. Merrell Jr., and T.A. Handley (eds.). Fish and wildlife relationships in old-growth forests. Proceedings of a symposium. 12-15 April 1982, Juneau, Alaska. American Institute of Fishery Research Biologists, Moorehead City, N.C. 425 pp.

Zarnowitz, J.E., and D.A. Manuwal. 1985. The effects of forest management on cavity-nesting birds in northwestern Washington. Journal of Wildlife Management 49(1):255-263.

8
Managing for Edge Effects
Joan Voller

Definitions

An *edge* is the interface between two types of habitat: the boundary where two communities meet (Giles 1978; Thomas et al. 1979; Forman and Godron 1986). An *ecotone* is the area of habitat affected by the meeting of the two communities (Thomas et al. 1979).

Edge effects: Edge effects were originally defined by Leopold (1933) as the tendency for boundaries between two habitat types to support a greater variety of species and number of individuals than either adjacent habitat. More recently, the term *edge effect* has come to describe the diverse phenomena that occur in the ecotone, and either positive or negative connotations may be associated with it (Angelstam 1992). For example, *edge effect* is now used to describe such phenomena as the modified environmental conditions found at the boundary between two habitats (Olympic Learning Center 1991; FEMAT 1993). In addition, the term may be used to describe the effects of edges on various processes, such as increased predation and parasitism of vulnerable species in the vicinity of the ecotone (Alverson et al. 1988).

Edge orientation: Edge orientation has recently been shown to play an important role for all variables of microclimate, and thus will affect vegetation growth and wildlife use (Chen et al. 1995). This is especially true in high-latitude areas such as British Columbia, where solar altitudes are high. Edge effects at high latitudes are strongest at southwest-facing edges and weakest at northeast-facing edges (Chen et al. 1995).

Inherent versus induced edge: There are two types of edge: inherent and induced (Yahner 1988). An inherent or natural edge is a long-term, natural feature of the landscape, such as the boundary between a riparian area and upland habitat. Inherent edges may be caused by topographic differences, changes in soil type, microclimate changes, or geomorphic features (Thomas et al. 1979). An induced edge is usually a relatively short-lived, often man-made feature such as that found between an old-growth forest and a clearcut (Yahner 1988). These edges may be caused by both natural and human disturbances such as fire, flooding, erosion, timber harvest, planting, or grazing (Thomas et al. 1979).

Abrupt versus feathered edge: Edges may also be described as 'abrupt' or 'feathered' (Ratti and Reese 1988). An abrupt edge shows a definite and sudden change between two habitat types. This type of edge is often an induced edge, as described above. A feathered edge shows a more gradual change from one habitat type to another. It may be inherent, such as that found along an elevational gradient, or induced, such as a forest/clearcut edge with only partial timber removal.

Edge contrast: the degree to which two adjacent habitats differ (Olympic Learning Center 1991), such as the contrast between a clearcut and a mature forest.

Depth of edge influence: Chen et al. (1992) describe the depth of edge influence (edge width) as the transition zone where the two systems interact.

An example of edge contrast between old growth, second growth, and a clearcut. (Photograph by Joan Voller)

This depth changes depending on which condition is being measured. For example, the depth of edge influence for wind is different from that for air temperature or humidity.

Edge species: Certain species dominate and flourish in the edge habitat (Yahner 1988). These species spend most of their time or are more abundant in the edge habitat and are classified as edge species (e.g., deer) (Nyberg and Janz 1990; Kremsater and Bunnell 1992).

Background

Aldo Leopold (1933) was one of the first to discuss the concept of edge. As part of his 'laws of dispersion and interspersion,' he stated that increased amounts of edge resulted in higher wildlife population densities. Leopold speculated that greater wildlife species diversity also occurred at the edge either because of the variety of vegetation between the two habitats or because of the availability of two adjacent habitats (Yahner 1988). His theory soon had managers creating edges to enhance wildlife habitat (Paton 1994). Scientists found that edge habitat appeared to enhance bird species diversity, although not necessarily in favour of indigenous species (Robinson 1988). This planned creation of edges also produced an overabundance of deer in the eastern United States (Alverson et al. 1988). Harrison (1984) stated that the 'greatest enemy' of the Kentucky Warbler (*Oporornis formosus*) was the white-tailed deer (*Odocoileus virginianus*) because it overgrazed the forest understorey upon which the warbler depended. Some researchers have stated that much of the evidence for species diversity in edge habitats has been circumstantial and has lacked the backing of strong scientific data (Reese and Ratti 1988).

Over the last two decades, scientists have recognized the limitations of Leopold's speculations; they have realized that this view took only certain edge species into consideration (Paton 1994). In addition, maximum abundance of game was no longer what managers strove for; naturally occurring species diversity now took precedence as a management objective (Yahner 1988). Scientists have also come to realize the association between fragmentation and increased edge habitat (Lehmkuhl and Ruggiero 1991); as fragmentation broke up the forest into patches, the amount of associated edge increased and interior habitat decreased (refer to Chapter 5 for more information on interior habitat, and Chapter 2 for information on fragmentation). Once 50% of the landscape was cut, the area that edge influenced decreased, but no interior habitat remained (Franklin and Forman 1987; Chen et al. 1995). This caused concern for species that seemed to require interior habitat or habitat away from open areas.

The decline in populations of neotropical migratory birds is often associated with this concern. A frequently cited explanation for this decline is increased nest predation and parasitism along edges (Paton 1994). Gates and Gysel (1978) developed the 'ecological trap' hypothesis, which stated that edges attracted a concentration of nests and thus created an increase in density-dependent mortality. Ratti and Reese (1988) and Yahner and Wright (1985) both argue that their studies disproved Gates and Gysel's (1978) theory and that no increase in nest predation consistent with the 'ecological trap' hypothesis was evident in their results. Angelstam (1986) and others (Ratti and Reese 1988; Yahner and Scott 1988) working with artificial nests stated that the relevance of their studies in relation to real nests is unknown. Ratti and Reese (1988) go on to argue that studies using real nests (e.g., Gates and Gysel 1978) are also potentially biased because of the variable nest detection rates between field and forest habitats. Studies using real nests (e.g., Chasko and Gates 1982) support Gates and Gysel's 'ecological trap' hypothesis, whereas studies using artificial nests (e.g., Angelstam 1986; Ratti and Reese 1988) fail to support it. Paton (1994), in his review and reanalysis of nest predation data, found strong evidence that nest success did decline near edges. The most conclusive studies he reanalyzed suggested that edge effects occurred within 50 m of the edge. He concluded that many of the discrepancies between studies resulted from poor and inconsistent experimental designs.

Many studies have looked at the depth to which edge effects penetrate the forest. Studies have shown that, depending on factors such as habitat and species, this may be anywhere from 15 m to 5 km (Laurance and Yensen 1991).

Ecological Principles

Edge Effects and Microclimatic and Vegetation Gradients

A microclimatic gradient exists between two adjacent habitat types; the intensity of this gradient is usually dependent on the structural differences between the patches (Olympic Learning Center 1991). When a boundary is between a forest and a clearcut, the modified environmental conditions on the forest floor consist of more light, stronger winds, and greater temperature and moisture variations (Chen et al. 1992). Generally, a clearcut receives more sun and wind, and is therefore warmer and drier than the adjacent forest in summer and wetter and cooler at night (Chen 1991). However, this pattern is strongly influenced by weather conditions. For example, on cloudy, rainy days there may not be any significant difference in microclimate between the two habitats, and on other days the forest may be warmer then the clearcut (Chen 1991). The conditions at the edge, however, are not always intermediate between the two habitats: solar radiation and wind velocity are intermediate, but during the day air temperature and relative humidity are most extreme along the edge (Chen 1991).

Edges between clearcuts and old growth have
distinct differences in microclimate, whereas edge
between old growth and a natural edge (in this
case slide vegetation) are less distinct. (Photograph
by Joan Voller)

The depth of edge influence for microclimatic conditions varies depending on the type of vegetation along the edge (shrubs versus trees), edge type
(feathered versus abrupt) (Angelstam 1992), orientation, and daily weather
conditions (Chen et al. 1990). Edge effects between a forest and a clearcut,
versus an older vegetated forest/agriculture edge, may go much deeper because of the sharpness of the edge (Angelstam 1992). Generally, the temperature and humidity will return to interior-like conditions 120-140 m
from an old-growth, Douglas-fir forest/clearcut edge (Chen et al. 1990).
However, during extreme heat and wind on a south exposure, the depth of
edge influence may penetrate more than 240 m into the stand (Chen et al.
1990). Both soil temperature and moisture exhibit a narrower depth of edge
influence (60-120 m). The radiation condition in the forest depends more

on forest structure and amount of exposed edge than on distance from the edge itself. However, on west- and east-facing slopes there is a deeper edge (120 m) than on south- and north-facing slopes (30-60 m). The depth of edge influence for wind is highly dependent on vegetation type, understorey, and stand density (Chen 1991). Chen (1991) found that in an old-growth Douglas-fir forest, it took up to 240 m for the wind to equal interior conditions.

A vegetation gradient from forest edges is thought to occur because of varying microclimatic conditions (Palik and Murphy 1990). In northern temperate areas, edges on the north aspect usually have a narrower vegetation gradient then those on southern aspects (Palik and Murphy 1990). Elevated tree growth and abundant regeneration of species such as Douglas-fir (*Pseudotsuga menziesii*) and western hemlock (*Tsuga heterophylla*) often occur in the first 30-60 m (Chen et al. 1992). In addition, higher tree mortality often occurs along edges (Chen et al. 1992). Chen et al. (1992) and others (Ruth and Yoder 1953; Gratowski 1956; Alexander 1964; Ranney 1977; Williams-Linera 1990) have also noted a higher number of snags and fallen trees near edges. Trees along edges of newly exposed clearcuts are susceptible to windfall. This is because support mechanisms needed to withstand strong winds were not needed during the trees' growth and therefore were not developed (Saunders et al. 1991).

Decomposition rates of fine litter and coarse woody debris were found to be higher near edges then in either adjacent forests or clearcuts (Edmonds and Bigger 1984). As well, Chen et al. (1995) observed that the depth of organics was much lower within 30 m of a southeast-facing edge of a Douglas-fir forest, suggesting a decomposition gradient going from the edge into the forest.

Edge Effects and Invertebrates

Geiger (1965) and Ranney (1977) found that trees adjacent to clearcuts often became stressed and, as a result, were more susceptible to insect infestation and disease. The increase in dead and fallen trees along the forest edge leads to increased habitat for invertebrates (Alexander 1964). Hansson (1983) noted that entomologists acknowledge that weakened edge trees often have a richer insect fauna than interior forest trees. Helle and Muona (1985) also found a strong preference by invertebrates for forest edge. Geiger (1965) noted that along with the altered temperature regimes along edges, invertebrate competition, predator-prey relationships, and parasitic relationships became destabilized. For example, the Vienna moth (*Ocneria monacha*) can emerge from its eggs laid on warm tree trunks along the forest edge earlier than its parasite, which emerges from the cooler forest floor; this head start can result in a fortified population of this pest (Geiger 1965).

The microclimatic differences along the edge (wind speed, temperature, humidity, and solar radiation) can significantly influence the migration and dispersal of flying insects. Wing beating for take-off and flight are influenced by air temperature, humidity, and light (Johnson 1969). Nutrient cycling may also be affected by the increase in soil temperature, and this in turn will affect the number and activity of invertebrates (Klein 1989; Parker 1989).

Edge Effects and Herpetofauna

Although there are few studies on amphibians and reptiles and edge per se, there are several studies on amphibians and reptiles in clearcuts versus forests. Results indicate that the difference between the two habitat types is a potential barrier to the movement of amphibians. Sadoway (1986) stated that moisture and temperature may be what most influences local distributions of amphibians. Heatwole (1962) also stated that temperature may be the most important factor influencing the microdistribution of plethodontid salamanders. Habitats that reach a higher temperature than an amphibian's critical thermal maxima will not be inhabited by that species (Sadoway 1986). At present this critical temperature is not known for most amphibian species in British Columbia (Sadoway 1986). Blymer and McGinnes (1977) stated that, because of temperature and soil moisture differences between the forest and clearcut, amphibians were unable to inhabit clearcuts for at least 8 years after logging. Bury (1983) found that amphibian species were still not present in clearcuts 6-14 years after logging.

Rosenberg and Raphael (1986) found that, overall, amphibian densities increased significantly in plots with more edge, but some species, such as the Pacific giant salamander (*Dicamptodon tenebrosus*), did not. However, they qualify their results by stating that fragmentation is a relatively new phenomenon in the Pacific Northwest and that the long-term effects may have yet to manifest themselves. Blaustein et al. (1994) argue that fragmentation and habitat loss are the major causes of amphibian declines. They reason that fragmentation is restricting the ability of amphibians to recolonize sites that have experienced local extinctions. Lehmkuhl and Ruggiero (1991) (through a model on the effects of fragmentation on certain species) determined that 12 of the 15 amphibian species they listed in the Pacific Northwest would be highly susceptible to extinction if their habitats became fragmented.

Reptiles may be found in a variety of habitats and, as desiccation is not a problem, are often found basking in clearcuts and other open areas (Sadoway 1986). The northwestern garter snake (*Thamnophis ordinoides*) favours the dense vegetation along the edge between early successional stages and open habitat (Gregory 1978). The northern alligator lizard (*Gerrhonotus coeruleus*) also uses edge habitat where dry, rocky areas lie adjacent to forests. Both species utilize edge habitat because it houses prey species and provides hiding areas and hibernacula (Sadoway 1986).

Edge Effects and Birds

Much of the original research on fragmentation and edge effects occurred in the eastern United States and focused on the declining migratory bird populations. The decline is often attributed to the increased predation and nest parasitism that can occur in conjunction with increased edge habitat (Paton 1994).

Leopold (1933) and Odum's (1971) original definition of edge effects appears to be especially applicable to bird populations. Many studies have demonstrated that avian species diversity is higher in edge habitat than in uniform habitat (Gates and Gysel 1978). Kroodsma (1982) found peaks in bird species density at the forest edge and again 217 m into the forest. Strelke and Dickson (1980) found that bird numbers, diversity, and abundance were highest in the first 25 m of the forest. Rosenberg and Raphael (1986) also found that bird species richness increased with increased fragmentation and edge. However, they point out two possible explanations for this increased species richness: rare or low-density species may not have been included in the study because of reduced sampling effort, and fragmentation in the Pacific Northwest is a relatively new phenomenon and long-term effects may not be evident yet. Whitcomb et al. (1981) found that some forest interior species increased initially when their habitat became fragmented, but over time they declined severely in numbers or became eliminated.

As with other groups, some bird species are positively affected by increases in edge habitat while others are negatively affected. Derleth et al. (1989) found that in general bird species richness and density increased along the forest edge; associated with this increase, however, was the loss of some forest interior species. Hansson (1983) stated that clearcutting may indirectly contribute to increased bird populations along edges by increasing insect populations. In his studies in Sweden, he found that the density of many tree-gleaning species increased along the forest/clearcut edge. However, in the northeastern United States, the loss of large tracts of interior habitat has seriously threatened neotropical migrants (Robbins 1979). Anderson (1979) found that although the overall number of species along the edge increased, the number of migrants decreased. However, Welsh (1987) found that several of the neotropical migrants threatened by fragmentation in the northeastern U.S. were common even in small habitat patches in the boreal forests of Ontario. He argued that it should not be assumed that a species would react in the same way to the same conditions throughout its range. His study also found that different species used different successional stages. In the U.S. Pacific Northwest, Rosenberg and Raphael (1986) found that resident birds and birds with large home ranges were most affected by fragmentation. Wilcove et al. (1986) stated that the increase in bird species diversity may be misleading because habitat generalists increase while habitat specialists decrease. In a study in Wyoming, Keller and Anderson (1992) found that the presence or absence of a bird species was more closely associated with the type of adjacent habitat rather than the existence of an edge.

Gates and Gysel (1978) found that the main causes of nest failure in their study were predation and nest parasitism. An increase in brood parasitism by the Brown-headed Cowbird (*Molothrus ater*) may explain recent declines in the songbirds of eastern forests (Brittingham and Temple 1983). Expanding habitat fragmentation and exposure of forest edge is blamed for this increase. Originally, the Brown-headed Cowbird was restricted to the plains and prairies of the central United States. Because of increases in open habitat brought about by agriculture and the widespread introduction of livestock, however, the Brown-headed Cowbird has managed to spread both eastward and westward to become a serious threat to many songbirds with inadequate defences against parasitism (Brittingham and Temple 1983). Because of its parasitic lifestyle, this species now occupies a wider range of habitats than any other passerine in North America (Robinson et al. 1992). The cowbird is now commonly found throughout many areas of British Columbia (Terres 1991).

Gates and Gysel (1978) found that cowbird parasitism increased and host fledgling success decreased near the forest edge. O'Conner and Faaborg (1992) found that cowbirds penetrated at least 400 m into the forest, but their abundance decreased significantly as distance from the edge increased. They found the same result with interior openings; cowbirds would use an opening as small as 0.2 ha, from which they could penetrate the interior forest. Brittingham and Temple (1983) found that cowbirds were just as likely to use a forest/clearcut edge as a forest/agriculture edge. The female cowbird lays an average of 30-40 eggs per season; therefore very few cowbirds are needed to parasitize several nests (Robinson et al. 1992).

It has been demonstrated that cowbirds have a negative effect on vulnerable species such as the Kirtlands Warbler (*Dendroica kirtlandii*). However, Trail and Baptista (1993) also demonstrated the devastating effects of cowbird parasitism on a widespread songbird such as the Nuttall's White-crowned Sparrow (*Zonotrichia leucophrys*). They found a parasitism rate of 40-50% in the San Francisco area, an increase from the 5% of 15 years before. They calculated that this species of sparrow could withstand only 20% parasitism before its long-term survival became threatened. Cowbird parasitism greater than 30% has been shown to cause destabilized populations that were vulnerable to local extirpations (Laymon 1987).

Many studies have looked at the effects of predation on nesting birds along field/forest edges, but few have looked at such effects along forest/ clearcut edges (Rudnicky and Hunter 1993). Angelstam (1986) found that predation pressure on artificial nests in fragmented habitats varied according to the degree of edge contrast between two adjacent habitats rather than according to the size of the patch. Rudnicky and Hunter (1993) also argue that relatively new abrupt edges between forests and clearcuts are likely to attract fewer birds and fewer predators because they offer less transition zone vegetation. They found no difference in predation on artificial ground nests between forest/clearcut edges and interior habitat. However, predation on nests in shrubs was higher along the edge than in either clearcuts or forest.

Yahner and Scott (1988) found that the greater the amount of fragmented habitat surrounding a patch, the greater the amount of nest predation. An area surrounded by 50% fragmented habitat had the highest predation, while an area with no surrounding fragmentation had the lowest. Other studies have found that edge-related predation pressure on artificial ground nests came chiefly from predators of the surrounding habitat (Andren and Angelstam 1988; Small and Hunter 1988; Yahner and Scott 1988). Ratti and Reese (1988) found that predation pressure was greater along an abrupt edge (e.g., forest/clearcut edge) than along a feathered edge (e.g., an edge with only partial timber removal). They stated that natural edges are less abrupt than forest/agriculture or forest/clearcut edges, and therefore offer greater protection from predation. Andren and Angelstam (1988) found that predation pressure decreased to levels found in continuous forests within 200-500 m of a forest/farmland edge. Wilcove et al. (1986) proposed that edge-related predation may penetrate the forest as much as 600 m.

Edge Effects and Mammals

Harris (1988) stated that, 'With a few exceptions, terrestrial mammals are highly vulnerable to forces that create impassable barriers between component habitats in the landscape.' Zuleta (1993) found a strong edge effect for small mammals in the lower Fraser Valley. He concluded that even the larger fragments would be unable to sustain viable populations because of the large amount of edge. For many mammals, these barriers may come in the form of an induced edge, such as along the side of a road (Harris 1988). In southeastern Ontario and Quebec, studies have shown that species such as the eastern chipmunk (*Tamias striatus*), gray squirrel (*Sciurus carolinensis*), and white-footed mouse (*Peromyscus leucopus*) rarely crossed roads that were wider than 20 m between two forest edges (Oxley et al., in Noss 1990). Merriam (in Noss 1990) also found that narrow gravel roads constituted barriers to white-footed mice. Even 'edge species' such as deer (*Odocoileus* spp.) and elk (*Cervus elaphus*) have shown an adverse response to roads

Edges can be an important component in the habitat matrix for species such as elk. (Photograph by Joan Voller)

(Kucera and McCarthy 1988). Lyon (1983) describes a model using three data sets that demonstrates an avoidance response by elk to traffic on forest roads. Other studies have also shown a decline in elk use of habitat adjacent to forest roads (Thomas et al. 1979). Rost and Bailey (1979) found that deer use at 300-400 m from the road was triple what it was 100 m from the road. On Vancouver Island, deer and elk were often observed near busy roads or using roads for travel. However, avoidance occurred if there was insufficient cover along the edge of the road (Nyberg and Janz 1990).

For some species, edge habitat such as that between forests and clearcuts means increased resources. Many ungulates thrive in edge habitats. Several studies have found significantly greater deer use along a forest/clearcut edge than in the interior of either habitat (Blymer and Mosby 1977; Clark and Gilbert 1982; Hanley 1983; Nyberg and Janz 1990; Kremsater and Bunnell 1992). In Wisconsin, timber harvest and the intentional creation of 'wildlife openings' have boosted the deer population to excessive numbers (more than 4 deer per square kilometre) (Alverson et al. 1988). The promotion of high numbers of edge species such as deer may change the balance and composition of the entire community (Alverson et al. 1988). A study in Wisconsin demonstrated that intensive deer browsing reduced the total plant species richness by one-quarter compared with a control area. In addition, moose (*Alces alces*) are thought to be excluded from some areas with excessive deer populations by the presence of deer parasites such as brainworm (*Parelaphostrongylus tenuis*) (Alverson et al. 1988).

It is suggested that leaving green trees dispersed throughout a clearcut may reduce edge contrast. (Photograph by Scott Harrison)

Managing for Edge Effects

Stand Level

Chen et al. (1995) suggested that a possible approach to minimize edge effects at the stand level was to create feathered edges instead of abrupt edges, and to preserve understorey vegetation. This would allow a higher resistance to physical and biological variables such as wind, moisture loss, excessive heat, and invasive wildlife. It would also allow a narrower depth of edge influence, thus increasing the amount of interior habitat (Chen 1991).

Ghuman and Lal (1987) and Franklin (1992) suggested that another approach might be to leave more green trees within the clearcut, especially near the edge, which in turn might reduce the contrast between the clearcut and the forest. Leaving more green trees within a clearcut also more closely mimics a moderate-intensity fire or wind disturbance (North et al. 1996). It would also reduce the depth of edge influence and thus preserve more interior habitat (Chen et al. 1995).

Oliver and Larson (1990) pointed out that as a forest regrows and the contrast between it and the adjacent forest is reduced, the edge effect is also reduced. Therefore, increasing the rate at which a clearcut is planted and accelerating tree growth would help speed up this transition.

As a forest regrows, the contrast between it and the adjacent forest is reduced, thus decreasing edge contrast. (Photograph by Joan Voller)

Landscape Level

Chen et al. (1995) suggested that a possible approach to minimize edge effects at the landscape level was to retain large forest patches instead of small ones. For example, if the depth of edge influence is 400 m, any forest patch smaller than 64 ha will have lost its interior habitat (Chen et al. 1995). As well, after 50% of the forest in a landscape is cut, no interior habitat will remain (Franklin and Forman 1987). Swanson et al. (1992) pointed out that the aggregation of harvest patches (versus the harvest of many small patches) would also reduce the amount of edge. Unfortunately, the resulting larger harvest patch might have an extremely harsh environment, which would counter the advantages of reduced edge.

Laurance and Yensen (1991) noted that the shape of a patch also influenced the amount of edge. A circular or square patch had the least amount of perimeter and therefore the least amount of edge. (For more information on patch shape, refer to Chapters 2 and 5.)

Research Needs

As forests have become more fragmented through timber harvest, agriculture, and urbanization, the amount of edge habitat has increased (Keller and Anderson 1992). Research on the effects of edges has led to doubt about the traditional belief that edges are an entirely positive feature in the landscape (Ratti and Reese 1988). However, most studies to date have been incompatible with each other with respect to factors such as experimental design, location, edge structure, edge type and age, species, and method of analysis (Ratti and Reese 1988; Reese and Ratti 1988; Paton 1994). The importance of understanding edge effects in relation to ecosystem structure and function, fragmentation, and wildlife habitat management requires that further research be undertaken, with emphasis on large sample sizes and replication (Ratti and Reese 1988). The following are the major information gaps identified in the literature:

The influence of edges at different spatial and temporal scales. To further our understanding of edge dynamics, researchers need to study edges at various spatial (e.g., patch versus landscape) and temporal (e.g., seasonal) scales (Chen et al. 1995).

Depth of edge influence. For landscape-level management of fragmented habitats to be effective, the role of edges must be understood for all types of forest. For example, in order to determine the amount of forest capable of supporting plant and animal species that represent natural ecosystem conditions, it is necessary to estimate the depth of edge influence for each forest type (Palik and Murphy 1990).

Soil and its related properties. Soil and its related properties have not been well studied in relation to edges. Recent ecological research has suggested that biogeochemical processes are key in understanding and predicting the changes in an ecosystem (J. Chen, pers. comm., 1996).

Distinctions between inherent and induced edges. Paton (1994) suggests the importance of studying the biological differences between natural and induced edges.

Studies of pre- and post-fragmented systems. These are necessary to understand the extent of edge effects and other processes on the internal environment (Saunders et al. 1991).

Edge phenomena at the landscape level. Further studies are needed to understand the effects of edges at the landscape level, e.g., edge phenomena in association with forest patch size, shape, and spatial configuration at multiple spatial and temporal scales (Angelstam 1992; Chen et al. 1992). With increasing habitat fragmentation and thus increasing amounts of edge habitat, more research is needed to understand the effects on such factors as location of species and predation, and the effects of fragmentation intensity on species (Santos and Telleria 1992). Long-term studies in a variety of landscapes are needed to understand the impacts of edges on communities, guilds, and 'key' species (Yahner 1988).

Amphibians and edges. Few studies have looked at the effect of edges on amphibians. Critical temperature maximums are unknown for most amphibians in B.C. (Sadoway 1986). This information, in association with studies on the depth of altered microclimatic conditions along all types and aspects of edges, is needed (Sadoway 1986; Chen et al. 1992).

Edges and bird nest predation and parasitism. Yahner et al. (1989) and Ratti and Reese (1988) recommend further study of nest predation along various types of edge and its effect on avian population numbers over time. Two angles still relatively unexplored are the relationship between nesting success and predator densities, and searching methods of nest predators (Paton 1994). Further research with sound experimental design at realistic landscape scales is needed to quantify the extent of nest predation and parasitism. More data are also needed on the potential threshold values of edge effects for nest success in a diversity of landscape designs and habitat types (Paton 1994). Few demographic studies have been undertaken to determine how much parasitism neotropical migrants can tolerate (Robinson et al. 1992).

Effects of forest management on bird populations. Little information is available on the effects of selective harvesting and uneven-aged management on bird populations – for example, how changes in stand structure and composition affect habitat suitability for forest birds, and how the creation of openings affects the rates of predation and cowbird parasitism on these species (Thompson 1993).

Edge effects and mammals. There is a need to better understand what resources species are responding to when they choose to use edges (Hanley 1983). Further research is also needed on the effects of fragmentation and edge on the persistence of wildlife populations over time and space; emphasis should be placed on the spatial dynamics of fragmentation at a landscape scale (Lehmkuhl and Ruggiero 1991).

Literature Cited

Alexander, R.R. 1964. Minimizing windfall around clear cuttings in spruce-fir forests. Forest Science 10:130-143.

Alverson, W.S., D.M. Waller, and S.L. Solheim. 1988. Forests too deer: edge effects in northern Wisconsin. Conservation Biology 2(4):348-358.

Anderson, S.H. 1979. Changes in forest bird species composition caused by transmission-line corridor cuts. American Birds 33(1):3-6.

Andren, H., and P. Angelstam. 1988. Elevated predation rates as an edge effect in habitat islands: experimental evidence. Ecology 69(2):544-547.

Angelstam, P. 1986. Predation on ground-nesting birds' nests in relation to predator densities and habitat edge. Oikos 47(3):365-373.

–. 1992. Conservation of communities – the importance of edges, surroundings and landscape mosaic structure. Chapter 2, pages 9-70 in Lennart Hansson (ed.). Ecological principles of nature conservation. Elsevier Applied Science, New York.

Blaustein, A.R., D.B. Wake, and W.P. Sousa. 1994. Amphibian declines: judging stability, persistence, and susceptibility of populations to local and global extinctions. Conservation Biology 8(1):60-71.

Blymer, M.J., and B.S. McGinnes. 1977. Observations on the possible detrimental effects of clearcutting on terrestrial amphibians. Bulletin of the Maryland Herpetological Society 13:79-83.

Blymer, M.J., and H.S. Mosby. 1977. Deer utilization of clearcuts in southwestern Virginia. Southern Journal of Applied Forestry 1:10-13.

Brittingham, M.C., and S.A. Temple. 1983. Have cowbirds caused forest songbirds to decline? BioScience 33(1):31-35.

Bury, B.R. 1983. Differences in amphibian populations in logged and old growth redwood forests. Northwest Science 57:167-178.

Chasko, G.G., and J.E. Gates. 1982. Avian habitat suitability along a transmission-line corridor in an oak-hickory forest region. Wildlife Monographs 82. 41 pp.

Chen, J. 1991. Edge effects: microclimatic pattern and biological responses in old-growth Douglas-fir forests. Ph.D. thesis, University of Washington, Seattle. 174 pp.

Chen, J., J.F. Franklin, and T.A. Spies. 1990. Microclimatic pattern and basic biological responses at the clearcut edges of old-growth Douglas-fir stands. Northwest Environmental Journal 6(2):424-425.

–. 1992. Vegetation responses to edge environments in old-growth Douglas-fir forests. Ecological Applications 2(4):387-396.

–. 1995. Growing-season microclimatic gradients from clearcut edges into old-growth Douglas-fir forests. Ecological Applications 5(1):74-86.

Clark, T.P., and F.F. Gilbert. 1982. Ecotones as a measure of deer habitat quality in central Ontario. Journal of Applied Ecology 19(3):751-758.

Derleth, E.L., D.G. McAuley, and T.J. Dwyer. 1989. Avian community response to small-scale habitat disturbance in Maine. Canadian Journal of Zoology 67(2):385-390.

Edmonds, R.L., and C.M. Bigger. 1984. Decomposition and nitrogen mineralization rates in Douglas-fir needles in relation to whole tree harvesting practices. Pages 187-192 in Proceedings of the 1983 Society of American Foresters national convention, Portland, Ore.

Forest Ecosystem Management Assessment Team (FEMAT). 1993. Forest ecosystem management: an ecological, economic, and social assessment. USDA Forest Service (and other agencies), Portland, Ore.

Forman, R.T.T., and M. Godron. 1986. Landscape ecology. John Wiley and Sons, New York.

Franklin, J.F. 1992. Scientific basis for new perspectives in forests and streams. Pages 25-72 in R.J. Naiman (ed.). Watershed management: balancing sustainability and environmental changes. Springer-Verlag, New York.

Franklin, J.F., and R.T.T. Forman. 1987. Creating landscape patterns by forest cutting: ecological consequences and principles. Landscape Ecology 1:5-18.

Gates, J.E., and L.W. Gysel. 1978. Avian nest dispersion and fledging success in field-forest ecotones. Ecology 59(5):871-883.

Geiger, R. 1965. The climate near the ground. Harvard University Press, Cambridge, Mass. 611 pp.

Ghuman, B.S., and R. Lal. 1987. Effects of partial clearing on microclimate in a humid tropical forest. Agricultural and Forest Meteorology 40:17-29.

Giles, R.H. Jr. 1978. Wildlife management. W.H. Freeman, San Francisco, Ca.

Gratowski, H.J. 1956. Windthrow around staggered settings in old-growth Douglas-fir. Forest Science 2:60-74.

Gregory, P.T. 1978. Feeding habits and diet overlap of three species of garter snakes (*Thamnophis*) on Vancouver Island. Canadian Journal of Zoology 56(9):1967-1974.

Hanley, T.A. 1983. Black-tailed deer, elk, and forest edge in a western Cascades watershed. Journal of Wildlife Management 47(1):237-242.

Hansson, L. 1983. Bird numbers across edges between mature conifer forest and clearcuts in Central Sweden. Ornis Scandinavica 14(2):97-103.

Harris, L.D. 1988. Edge effects and conservation of biotic diversity. Conservation Biology 2(4):330-332.

Harrison, H. 1984. Wood warbler's world. Simon and Schuster, New York. 335 pp.

Heatwole, H. 1962. Environmental factors influencing local distribution and activity of the salamander, *Plethodon cinereus*. Ecology 43(3):460-472.

Helle, P., and J. Muona. 1985. Invertebrate numbers in edges between clear-fellings and mature forests in northern Finland. Silva Fennica 19(3):281-294.

Johnson, C.G. 1969. Migration and dispersal of insects by flight. Methuen, London, England.

Keller, M.E., and S.H. Anderson. 1992. Avian use of habitat configurations created by forest cutting in southeastern Wyoming. Condor 94:55-65.

Klein, B.C. 1989. Effects of forest fragmentation on dung and carrion beetle communities in Central Amazonia. Ecology 70:1715-1725.

Kremsater, L.L., and F.L. Bunnell. 1992. Testing responses to forest edges: the example of black-tailed deer. Canadian Journal of Zoology 70:2426-2435.

Kroodsma, R.L. 1982. Edge effect on breeding forest birds along a power-line corridor. Journal of Applied Ecology 19:361-370.

Kucera, T.E., and C. McCarthy. 1988. Habitat fragmentation and mule deer migration corridors: a need for evaluation. Transactions of the Western Section of the Wildlife Society 24: 61-67.

Laurance, W.F., and E. Yensen. 1991. Predicting the impacts of edge effects in fragmented habitats. Biological Conservation 55:77-92.

Laymon, S.A. 1987. Brown-headed cowbirds in California: historical perspectives and management opportunities in riparian habitats. Western Birds 18:63-70.

Lehmkuhl, J.F., and L.F. Ruggiero. 1991. Forest fragmentation in the Pacific Northwest and its potential effects on wildlife. Pages 35-46 in L.F. Ruggiero, K.B. Aubry, A.B. Carey, and M.H. Huff (eds.). Wildlife and vegetation of unmanaged Douglas-fir forests. USDA Forest Service General Technical Report PNW-GTR-285. Portland, Ore.

Leopold, A. 1933. Game management. Charles Scribner's Sons, New York.

Lyon, L.J. 1983. Road density models describing habitat effectiveness for elk. Journal of Forestry 81(9):592-595, 613.

North, M., J. Chen, G. Smith, L. Krakowiak, and J. Franklin. 1996. Initial response of understory plant diversity and overstory tree diameter growth to a green tree retention harvest. Northwest Science 70(1):24-35.

Noss, R. 1990. The ecological effects of roads (or the road to destruction). Pages 1-5 in J. Davis (ed.). Killing roads: a citizen's primer on the effects and removal of roads. Earth First! Biodiversity Project Special Publication. Tucson, Ariz.

Nyberg, J.B., and D.W. Janz. 1990. Deer and elk habitats in coastal forests of southern British Columbia. B.C. Ministry of Forests, Special Report Series 5. Co-published by B.C. Ministry of Environment in cooperation with Wildlife Habitat Canada. 310 pp.

O'Conner, R.J., and J. Faaborg. 1992. The relative abundance of the brown-headed cowbird (*Molothrus ater*) in relation to exterior and interior edges in forests of Missouri. Transactions of the Missouri Academy of Science 26:1-9.

Odum, E.P. 1971. Fundamentals of ecology. W.B. Saunders, Philadelphia, Pa.

Oliver, C.D., and B.C. Larson. 1990. Forest stand dynamics. McGraw-Hill, New York.

Olympic Learning Center. 1991. Concepts in ecosystem management. Pamphlet series. Olympic Learning Center, Olympic National Forest, Quilcene, Wash.

Palik, B.J., and P.G. Murphy. 1990. Disturbance versus edge effects in sugar-maple/beech forest fragments. Forestry and Ecological Management 32:187-202.

Parker, C.A. 1989. Soil biota and plants in the rehabilitation of degraded agricultural soils. Pages 423-438 in J.D. Majer (ed.). Animals in primary succession. The role of fauna in reclaimed lands. Cambridge University Press, Cambridge, England.

Paton, P.W.C. 1994. The effect of edge on avian nest success: how strong is the evidence? Conservation Biology 8(1):17-26.

Ranney, J.W. 1977. Forest-island edges – their structure, development and importance to regional forest ecosystem dynamics. Environmental Sciences Division Publication No. 1069. Oak Ridge National Laboratory, Oak Ridge, Tenn.

Ratti, J.T., and K.P. Reese. 1988. Preliminary test of the ecological trap hypothesis. Journal of Wildlife Management 52(3):484-491.

Reese, K.P., and J.T. Ratti. 1988. Edge effect: a concept under scrutiny. Pages 127-136 in R.E. McCabe (ed.). New approaches in managing natural resources. Transactions of the 53rd

North American Wildlife and Natural Resources Conference. Wildlife Management Institute, Washington, D.C.

Robbins, C.S. 1979. Effect of forest fragmentation on bird populations. Pages 198-212 in R.M. DeGraaf and K.E. Evans (eds.). Management of north central and northeastern forests for nongame birds: workshop proceedings. 23-25 January 1979, Minneapolis, Minn. USDA Forest Service General Technical Report NC-51.

Robinson, S.K. 1988. Reappraisal of the costs and benefits of habitat heterogeneity for nongame wildlife. Pages 145-155 in R.E. McCabe (ed.). New approaches in managing natural resources. Transactions of the 53rd North American Wildlife and Natural Resources Conference. Wildlife Management Institute, Washington, D.C.

Robinson, S.K., J.A. Grzybowski, S.I. Rothstein, M.C. Brittingham, L.J. Petit, and F.R. Thompson. 1992. Management implications of cowbird parasitism on neotropical migrant songbirds. Pages 93-102 in D. Finch and P.W. Stangle (eds.). Status and management of neotropical migratory birds. USDA Forest Service General Technical Report RM-229.

Rosenberg, K.V., and M.G. Raphael. 1986. Effects of forest fragmentation on vertebrates in Douglas-fir forests. Pages 263-272 in J. Verner, M.L. Morrison, and C.J. Ralph (eds.). Wildlife 2000: modeling habitat relationships of terrestrial vertebrates. University of Wisconsin Press, Madison, Wis.

Rost, G.R., and J.A. Bailey. 1979. Distribution of mule deer and elk in relation to roads. Journal of Wildlife Management 43:634-641.

Rudnicky, T.C., and M.L. Hunter Jr. 1993. Avian nest predation in clearcut, forests, and edges in a forest-dominated landscape. Journal of Wildlife Management 57(2):358-364.

Ruth, R.H., and R.A. Yoder. 1953. Reducing wind damage in the forests of the Oregon Coast Range. USDA Forest Service PNW Research Paper 7.

Sadoway, K.L. 1986. Effects of intensive forest management on amphibians and reptiles of Vancouver Island: problem analysis. B.C. Ministry of Forests, Victoria, B.C. IWIFR-23:42 pp.

Santos, T., and J.L. Telleria. 1992. Edge effects on nest predation in Mediterranean fragmented forests. Biological Conservation 60:1-5.

Saunders, D.A., R.J. Hobbs, and C.R. Margules. 1991. Biological consequences of ecosystem fragmentation: a review. Conservation Biology 5(1):18-32.

Small, M.F., and M.L. Hunter. 1988. Forest fragmentation and avian nest predation in forested landscapes. Oecologia 76:62-64.

Strelke, W.K., and J.G. Dickson. 1980. Effect of forest clear-cut edge on breeding birds in east Texas. Journal of Wildlife Management 44(3):559-567.

Swanson, F.J., R.P. Neilson, and G.E. Grant. 1992. Some emerging issues in watershed management: landscape patterns, species conservation, and climate change. Pages 307-323 in R.J. Naiman (ed.). Watershed management: balancing sustainability and environmental changes. Springer-Verlag, New York.

Terres, J.K. 1991. The Audubon Society encyclopedia of North American birds. Wing Books, New York. 1,109 pp.

Thomas, J.W., C. Maser, and J.E. Rodiek. 1979. Edges. Chapter 4 in J.W. Thomas (tech. ed.). Wildlife habitats in managed forests: the Blue Mountains of Oregon and Washington. USDA Agricultural Handbook No. 553. 512 pp.

Thompson, F.R. 1993. Simulated responses of a forest-interior bird population to forest management options in central hardwood forests of the United States. Conservation Biology 7(2):325-333.

Trail, P.W., and L.F. Baptista. 1993. The impact of brown-headed cowbird parasitism on populations of the Nuttall's White-crowned Sparrow. Conservation Biology 7(2):309-315.

Welsh, D.A. 1987. The influence of forest harvesting on mixed coniferous-deciduous boreal bird communities in Ontario, Canada. Acta Oecologica. Oecologia generalis 8(2):247-252.

Whitcomb, R.F., C.S. Robbins, J.F. Lynch, B.L. Whitcomb, M.K. Klimkiewicz, and D. Bystrak. 1981. Chapter 8, pages 125-204 in R.L. Burgess and D.M. Sharpe (eds.). Forest island dynamics in man-dominated landscapes. Springer-Verlag, New York. 310 pp.

Wilcove, D.S., C. McLellan, and A. Dobson. 1986. Habitat fragmentation in the temperate zone. Pages 237-256 in M.E. Soulé (ed.). Conservation biology: the science of scarcity and diversity. Sinauer Associates, Sunderland, Mass.

Williams-Linera, G. 1990. Vegetation structure and environmental conditions of forest edges in Panama. Journal of Ecology 78:356-373.

Yahner, R.H. 1988. Changes in wildlife communities near edges. Conservation Biology 2(4):333-339.

Yahner, R.H., and D.P. Scott. 1988. Effects of forest fragmentation on depredation of artificial nests. Journal of Wildlife Management 52(1):158-161.

Yahner, R.H., and A.L. Wright. 1985. Depredation on artificial ground nests: effects of edge and plot age. Journal of Wildlife Management 49:508-513.

Yahner, R.H., T.E. Morrell, and J.S. Rachael. 1989. Effects of edge contrast on depredation of artificial avian nests. Journal of Wildlife Management 53(4):1135-1138.

Zuleta, G.A. 1993. Analysis of habitat fragmentation effects with emphasis on small mammals at risk. Unpublished manuscript, UBC Ecology Group, Zoology Department, University of British Columbia, Vancouver, B.C. 24 pp.

Index